Florian Ion PETRESCU &
Relly Victoria PETRESCU

GEAR
SOLUTIONS

USA 2011

Scientific reviewer:

Dr. Veturia CHIROIU
Honorific member of
Technical Sciences Academy of Romania (ASTR)
PhD supervisor in Mechanical Engineering

Copyright

Title book: Gear Solutions

Authors book: Florian Ion PETRESCU & Relly Victoria PETRESCU

ISBN 978-1-4679-8764-6

WELCOME

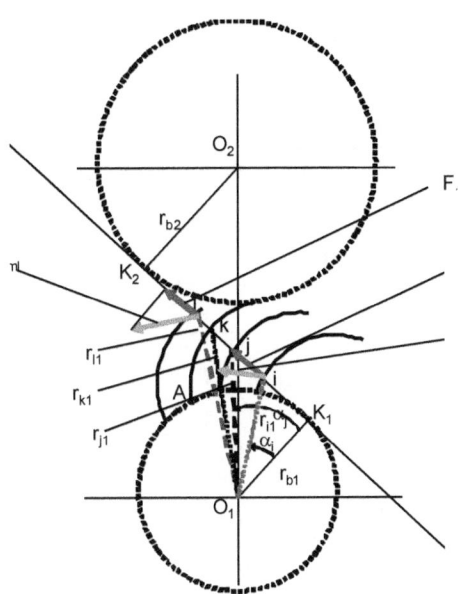

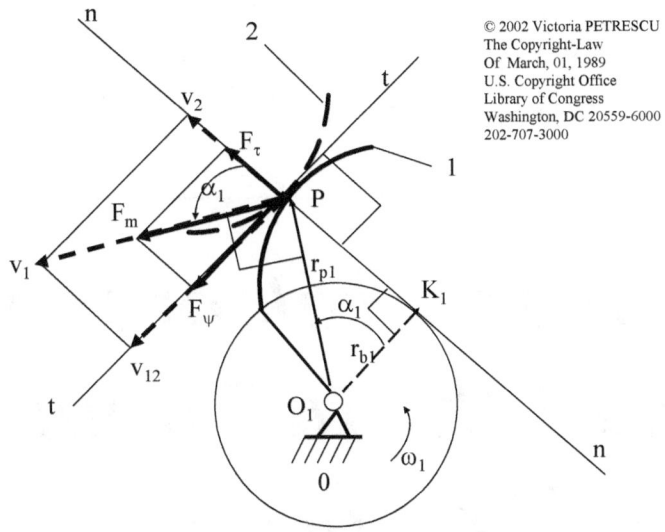

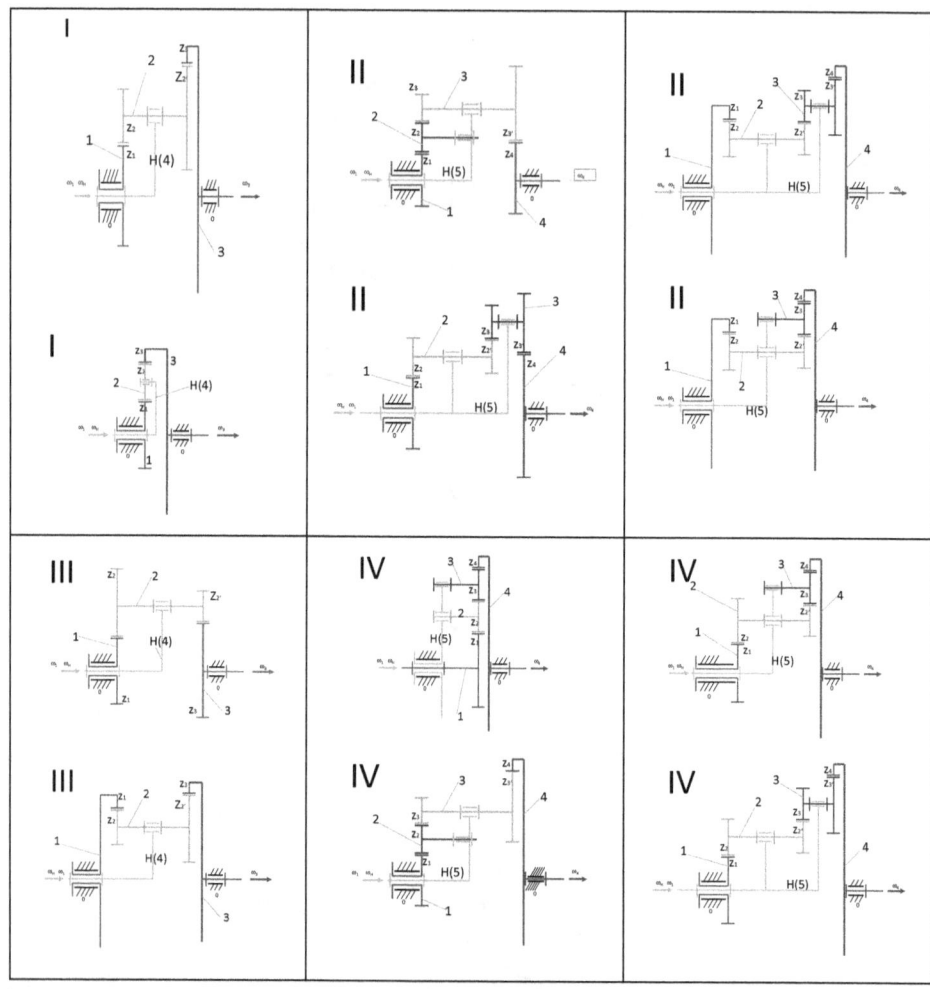

You are welcome to
read the full book!
The authors.

CHAPTER I

FORCES AND EFFICIENCY OF GEARING

ABSTRACT: *The chapter presents an original method to determine the efficiency of the gear, the forces of the gearing, the velocities and the powers. The originality of this method relies on the eliminated friction modulus. The chapter is analyzing the influence of a few parameters concerning gear efficiency. These parameters are: z_1 - the number of teeth for the primary wheel of gear; z_2 - the number of teeth of the secondary wheel of gear; α_0 - the normal pressure angle on the divided circle; β - the inclination angle. With the relations presented in this paper, it can synthesize the gear's mechanisms. Today, the gears are present everywhere, in the mechanical's world (In vehicle's industries, in electronics and electro-technique equipments, in energetically industries, etc...). Optimizing this mechanism (the gears mechanism), we can improve the functionality of the transmissions with gears.*

Keywords: Efficiency, forces, powers, velocities, gear, constructive parameters, teeth, outside circle, wheel.

1. INTRODUCTION

In this chapter the authors present an original method for calculating the efficiency of the gear, the forces of the gearing, the velocities and the powers.

The originality consists in the way of determination of the gear efficiency because it hasn't used the friction forces of couple (this new way eliminates the classical method).

It eliminates the necessity of determining the friction coefficients by different experimental methods as well.

The efficiency determinates by the new method is the same like the classical efficiency, namely the mechanical efficiency of the gear.

Precisely one determines the dynamics efficiency, but at the gears transmissions, the dynamics efficiency is the same like the mechanical efficiency; this is a greater advantage of the gears transmissions.

2. DETERMINING THE MOMENTARY DYNAMIC (MECHANICAL) EFFICIENCY, THE FORCES OF THE GEARING, AND THE VELOCITIES

The calculating relations [2, 3], are the next (1-21), (see the fig. 1):

$$\begin{cases} F_\tau = F_m \cdot \cos\alpha_1 \\ F_\psi = F_m \cdot \sin\alpha_1 \\ v_2 = v_1 \cdot \cos\alpha_1 \\ v_{12} = v_1 \cdot \sin\alpha_1 \\ \overline{F}_m = \overline{F}_\tau + \overline{F}_\psi \\ \overline{v}_1 = \overline{v}_2 + \overline{v}_{12} \end{cases} \qquad (1)$$

with: F_m - the motive force (the driving force);

$F\tau$ - the transmitted force (the useful force);

$F\psi$ - the slide force (the lost force);

v_1 - the velocity of element 1, or the speed of wheel 1 (the driving wheel);

v_2 - the velocity of element 2, or the speed of wheel 2 (the driven wheel);

v_{12} - the relative speed of the wheel 1 in relation with the wheel 2 (this is a sliding speed).

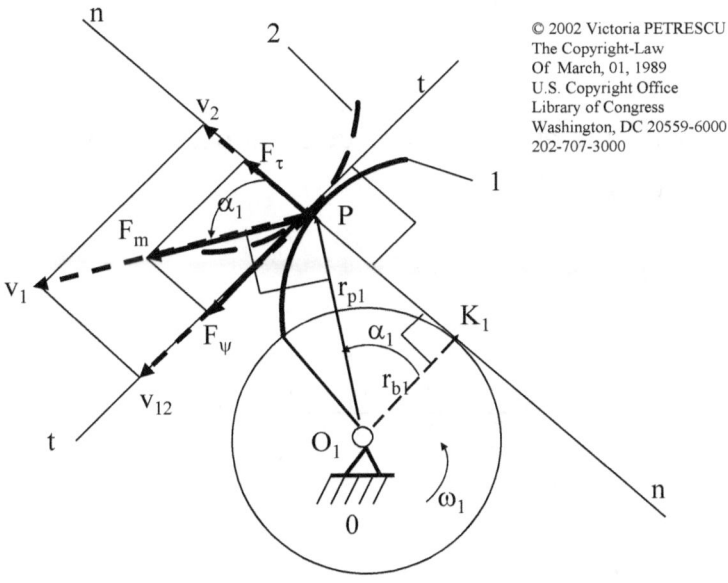

Fig. 1. *The forces and the velocities of the gearing*

The consumed power (in this case the driving power):

$$P_c \equiv P_m = F_m \cdot v_1 \qquad (2)$$

7

The useful power (the transmitted power from the profile 1 to the profile 2) will be written:

$$P_u \equiv P_\tau = F_\tau \cdot v_2 = F_m \cdot v_1 \cdot \cos^2 \alpha_1 \qquad (3)$$

The lost power will be written:

$$P_\psi = F_\psi \cdot v_{12} = F_m \cdot v_1 \cdot \sin^2 \alpha_1 \qquad (4)$$

The momentary efficiency of couple will be calculated directly with the next relation:

$$\begin{cases} \eta_i = \dfrac{P_u}{P_c} \equiv \dfrac{P_\tau}{P_m} = \dfrac{F_m \cdot v_1 \cdot \cos^2 \alpha_1}{F_m \cdot v_1} \\ \eta_i = \cos^2 \alpha_1 \end{cases} \qquad (5)$$

The momentary losing coefficient [1], will be written:

$$\begin{cases} \psi_i = \dfrac{P_\psi}{P_m} = \dfrac{F_m \cdot v_1 \cdot \sin^2 \alpha_1}{F_m \cdot v_1} = \sin^2 \alpha_1 \\ \eta_i + \psi_i = \cos^2 \alpha_1 + \sin^2 \alpha_1 = 1 \end{cases} \qquad (6)$$

It can easily see that the sum of the momentary efficiency and the momentary losing coefficient is 1:

Now one can determine the geometrical elements of gear. These elements will be used in determining the couple efficiency, η.

3. THE GEOMETRICAL ELEMENTS OF THE GEAR

We can determine the next geometrical elements of the external gear, [2,3], (for the right teeth, $\beta=0$):

The radius of the basic circle of wheel 1 (of the driving wheel), (7):

$$r_{b1} = \frac{1}{2} \cdot m \cdot z_1 \cdot \cos \alpha_0 \qquad (7)$$

The radius of the outside circle of wheel 1 (8):

$$r_{a1} = \frac{1}{2} \cdot (m \cdot z_1 + 2 \cdot m) = \frac{m}{2} \cdot (z_1 + 2) \qquad (8)$$

It determines now the maximum pressure angle of the gear (9):

$$\cos \alpha_{1M} = \frac{r_{b1}}{r_{a1}} = \frac{\frac{1}{2} \cdot m \cdot z_1 \cdot \cos \alpha_0}{\frac{1}{2} \cdot m \cdot (z_1 + 2)} = \frac{z_1 \cdot \cos \alpha_0}{z_1 + 2} \qquad (9)$$

And now one determines the same parameters for the wheel 2, the radius of basic circle (10) and the radius of the outside circle (11) for the wheel 2:

$$r_{b2} = \frac{1}{2} \cdot m \cdot z_2 \cdot \cos \alpha_0 \qquad (10)$$

$$r_{a2} = \frac{m}{2} \cdot (z_2 + 2) \qquad (11)$$

Now it can determine the minimum pressure angle of the external gear (12, 13):

$$\begin{cases} tg\alpha_{1m} = \dfrac{N}{r_{b1}} \\[2mm] N = (r_{b1} + r_{b2}) \cdot tg\alpha_0 - \sqrt{r_{a2}^2 - r_{b2}^2} = \\[2mm] = \dfrac{1}{2} \cdot m \cdot (z_1 + z_2) \cdot \sin\alpha_0 - \dfrac{m}{2} \cdot \sqrt{(z_2 + 2)^2 - z_2^2 \cdot \cos^2\alpha_0} = \\[2mm] = \dfrac{m}{2} \cdot [(z_1 + z_2) \cdot \sin\alpha_0 - \sqrt{z_2^2 \cdot \sin^2\alpha_0 + 4 \cdot z_2 + 4}] \end{cases} \qquad (12)$$

$$tg\alpha_{1m} = [(z_1 + z_2) \cdot \sin\alpha_0 - \sqrt{z_2^2 \cdot \sin^2\alpha_0 + 4 \cdot z_2 + 4}]/(z_1 \cdot \cos\alpha_0) \qquad (13)$$

Now we can determine, for the external gear, the minimum (13) and the maximum (9) pressure angle for the right teeth. For the external gear with bended teeth ($\beta \neq 0$) it uses the relations (14, 15 and 16):

$$tg\alpha_t = \dfrac{tg\alpha_0}{\cos\beta} \qquad (14)$$

$$tg\alpha_{1m} = [(z_1 + z_2) \cdot \dfrac{\sin\alpha_t}{\cos\beta} - \sqrt{z_2^2 \cdot \dfrac{\sin^2\alpha_t}{\cos^2\beta} + 4 \cdot \dfrac{z_2}{\cos\beta} + 4}] \cdot \dfrac{\cos\beta}{z_1 \cdot \cos\alpha_t} \qquad (15)$$

$$\cos\alpha_{1M} = \dfrac{\dfrac{z_1 \cdot \cos\alpha_t}{\cos\beta}}{\dfrac{z_1}{\cos\beta} + 2} \qquad (16)$$

For the internal gear with bended teeth ($\beta \neq 0$) it uses the relations (14 with 17, 18-A, or with 19, 20-B):

A. <u>When the driving wheel 1, has external teeth:</u>

$$tg\alpha_{1m} = [(z_1 - z_2) \cdot \frac{\sin\alpha_t}{\cos\beta} + \sqrt{z_2^2 \cdot \frac{\sin^2\alpha_t}{\cos^2\beta} - 4 \cdot \frac{z_2}{\cos\beta} + 4}] \cdot \frac{\cos\beta}{z_1 \cdot \cos\alpha_t} \qquad (17)$$

$$\cos\alpha_{1M} = \frac{\dfrac{z_1 \cdot \cos\alpha_t}{\cos\beta}}{\dfrac{z_1}{\cos\beta} + 2} \qquad (18)$$

B. <u>When the driving wheel 1, have internal teeth:</u>

$$tg\alpha_{1M} = [(z_1 - z_2) \cdot \frac{\sin\alpha_t}{\cos\beta} + \sqrt{z_2^2 \cdot \frac{\sin^2\alpha_t}{\cos^2\beta} + 4 \cdot \frac{z_2}{\cos\beta} + 4}] \cdot \frac{\cos\beta}{z_1 \cdot \cos\alpha_t} \qquad (19)$$

$$\cos\alpha_{1m} = \frac{\dfrac{z_1 \cdot \cos\alpha_t}{\cos\beta}}{\dfrac{z_1}{\cos\beta} - 2} \qquad (20)$$

4. DETERMINING THE EFFICIENCY

The efficiency of the gear will be calculated through the integration of momentary efficiency on all sections of gearing movement, namely from the minimum pressure angle to the maximum pressure angle, the relation (21), [2, 3].

$$\eta = \frac{1}{\Delta\alpha} \cdot \int_{\alpha_m}^{\alpha_M} \eta_i \cdot d\alpha = \frac{1}{\Delta\alpha} \int_{\alpha_m}^{\alpha_M} \cos^2 \alpha \cdot d\alpha = \frac{1}{2 \cdot \Delta\alpha} \cdot [\frac{1}{2} \cdot \sin(2 \cdot \alpha) + \alpha]_{\alpha_m}^{\alpha_M} =$$

$$= \frac{1}{2 \cdot \Delta\alpha}[\frac{\sin(2\alpha_M) - \sin(2\alpha_m)}{2} + \Delta\alpha] = \frac{\sin(2 \cdot \alpha_M) - \sin(2 \cdot \alpha_m)}{4 \cdot (\alpha_M - \alpha_m)} + 0.5$$

(21)

Table 1. Determining the efficiency of the gear's right teeth for $i_{12effective}= -4$

$i_{12effective}= -4$	right teeth					
$z_1 = 8$	$z_2 = 32$			$z_1 = 30$	$z_2 = 120$	
$\alpha_0 = 20^0$?	$\alpha_0 = 29^0$	$\alpha_0 = 35^0$		$\alpha_0 = 15^0$	$\alpha_0 = 20^0$	$\alpha_0 = 30^0$
$\alpha_m = -16.22^0$?	$\alpha_m = 0.7159^0$	$\alpha_m =$		$\alpha_m = 1.5066^0$	$\alpha_m = 9.5367^0$	α_m
$\alpha_M = 41.2574^0$	$\alpha_M = 45.5974$	$\alpha_M = 49.056$		$\alpha_M = 25.1018^0$	$\alpha_M = 28.241$	$\alpha_M = 35.718$
	$\eta = 0.8111$	$\eta = 0.7308$		$\eta = 0.9345$	$\eta = 0.8882$	$\eta = 0.7566$
$z_1 = 10$	$z_2 = 40$			$z_1 = 90$	$z_2 = 360$	
$\alpha_0 = 20^0$?	$\alpha_0 = 26^0$	$\alpha_0 = 30^0$		$\alpha_0 = 8^0$?	$\alpha_0 = 9^0$	$\alpha_0 = 20^0$
$\alpha_m = -9.89^0$?	$\alpha_m = 1.3077^0$	$\alpha_m =$		$\alpha_m = -0.1638^0$	$\alpha_m = 1.5838^0$	α_m
$\alpha_M = 38.4568^0$	$\alpha_M = 41.4966$	$\alpha_M = 43.806$		$\alpha_M = 14.3637^0$	$\alpha_M = 14.935$	$\alpha_M = 23.181$
	$\eta = 0.8375$	$\eta = 0.7882$			$\eta = 0.9750$	$\eta = 0.8839$
$z_1 = 18$	$z_2 = 72$					
$\alpha_0 = 19^0$	$\alpha_0 = 20^0$	$\alpha_0 = 30^0$				
$\alpha_m = 0.9860^0$	$\alpha_m = 2.7358^0$	α_m				
$\alpha_M = 31.6830^0$	$\alpha_M = 32.2505$	$\alpha_M = 38.792$				
$\eta = 0.90105$	$\eta = 0.8918$	$\eta = 0.7660$				

5. THE CALCULATED EFFICIENCY OF THE GEAR

We shall now see four tables with the calculated efficiency depending on the input parameters and once we proceed with the results we will draw some conclusions.

The input parameters are:

z_1 = the number of teeth for the driving wheel 1;

z_2 = the number of teeth for the driven wheel 2, or the ratio of transmission, i (i_{12}=-z_2/z_1);

α_0 = the pressure angle normal on the divided circle;

β = the bend angle.

Table 2. Determining the efficiency of the gear's right teeth for $i_{12effective}$= -2

$i_{12effective}$=-2	right teeth					
z_1=8	z_2=16			z_1=18	z_2=36	
α_0=20° ?	α_0=28°	α_0=35°		α_0=18°	α_0=20°	α_0=30°
α_m=-12.65° ?	α_m=0.9149°	α_m		α_m=0.6756°	α_m=3.9233°	α_m=18.6935°
α_M=41.2574°	α_M=45.0606	α_M=49.0559		α_M=31.1351°	α_M=32.250	α_M=38.7922°
	η=0.8141	η=0.7236		η=0.9052	η=0.8874	η=0.7633
z_1=10	z_2=20			z_1=90	z_2=180	
α_0=20° ?	α_0=25°	α_0=30°		α_0=8°	α_0=20°	α_0=30°
α_m=-7.13° ?	α_m=1.3330°	α_m=9.4106°		α_m=0.5227°	α_m	α_m=27.7825°
α_M=38.4568°	α_M=40.9522	α_M=43.8060		α_M=14.3637°	α_M=23.181	α_M=32.0917°
	η=0.8411	η=0.7817		η=0.9785	η=0.8836	η=0.7507

Table 3. Determining the efficiency of the gear's right teeth for $i_{12effective}= -6$

$i_{12effective}= - 6$	right teeth				
$z_1=8$	$z_2=48$		$z_1=18$	$z_2=108$	
$\alpha_0=20^0$?	$\alpha_0=30^0$	$\alpha_0=35^0$	$\alpha_0=19^0$	$\alpha_0=20^0$	$\alpha_0=30^0$
$\alpha_m=-17.86^0$?	$\alpha_m= 1.7784^0$	$\alpha_m=10.660^0$	$\alpha_m=0.4294^0$	$\alpha_m=2.2449^0$	α_m
$\alpha_M=41.2574^0$	$\alpha_M=46.1462^0$	$\alpha_M=49.0559$	$\alpha_M=31.6830^0$	$\alpha_M=32.2505^0$	$\alpha_M=38.792$
	$\eta=0.8026$	$\eta=0.7337$	$\eta=0.9028$	$\eta=0.8935$	$\eta=0.7670$
$z_1=10$	$z_2=60$		$z_1=90$	$z_2=540$	
$\alpha_0=20^0$?	$\alpha_0=26^0$	$\alpha_0=30^0$	$\alpha_0=9^0$	$\alpha_0=20^0$	$\alpha_0=30^0$
$\alpha_m=-11.12^0$?	$\alpha_m=0.6054^0$	$\alpha_m=$	$\alpha_m=1.3645^0$	$\alpha_m=16.4763^0$	α_m
$\alpha_M=38.4568^0$	$\alpha_M=41.4966^0$	$\alpha_M=43.8060$	$\alpha_M=14.9354^0$	$\alpha_M=23.1812^0$	$\alpha_M=32.091$
	$\eta=0.8403$	$\eta=0.7908$	$\eta=0.9754$	$\eta=0.8841$	$\eta=0.7509$

We begin with the right teeth (the toothed gear), with i=-4, once for z_1 we shall take successively different values, rising from 8 teeth. It can see that for 8 teeth of the driving wheel the standard pressure angle, $\alpha_0=20^0$, is so small to be used (it obtains a minimum pressure angle, α_m, negative and this fact is not admitted!). In the second table we shall diminish (in module) the value for the ratio of transmission, i, from 4 to 2. It will see how for a lower value of the number of teeth of the wheel 1, the standard pressure angle ($\alpha_0=20^0$) is to small and it will be necessary to increase it to a minimum value. For example, if $z_1=8$, the necessary minimum value is $\alpha_0=29^0$ for i=-4 (see the table 1) and $\alpha_0=28^0$ for i=-2 (see the table 2). If $z_1=10$, the necessary minimum pressure angle is $\alpha_0=26^0$ for i=-4 (see the table 1) and $\alpha_0=25^0$ for i=-2 (see the table 2).

When the number of teeth of the wheel 1 increases, it can decrease the normal pressure angle, α_0. One shall see that for $z_1=90$ it can take less for the normal pressure angle (for the pressure angle of reference), $\alpha_0=8^0$. In the table 3 it increases the module of i, value (for the ratio of transmission), from 2 to 6.

In the table 4, the teeth are bended ($\beta \neq 0$). The module i, take now the value 2.

Table 4. The determination of the gear's parameters in bend teeth for i=-4

$i_{12effective}=-4$	bend teeth	Table 4				
	$\beta=15^0$					
$z_1=8$	$z_2=32$			$z_1=30$	$z_2=120$	
$\alpha_0=20^0$?	$\alpha_0=30^0$	$\alpha_0=35^0$		$\alpha_0=15^0$	$\alpha_0=20^0$	$\alpha_0=30^0$
$\alpha_m=-16.836^0$	$\alpha_m=1.1265^0$	$\alpha_m=9.4455^0$		$\alpha_m=1.0269^0$	$\alpha_m=8.8602^0$	$\alpha_m=22.1550^0$
$\alpha_M=41.0834^0$	$\alpha_M=46.2592^0$	$\alpha_M=49.2953^0$		$\alpha_M=25.1344^0$	$\alpha_M=28.4591^0$	$\alpha_M=36.2518^0$
	$\eta=0.8046$	$\eta=0.7390$		$\eta=0.9357$	$\eta=0.8899$	$\eta=0.7593$
$z_1=18$	$z_2=72$			$z_1=90$	$z_2=360$	
$\alpha_0=19^0$	$\alpha_0=20^0$	$\alpha_0=30^0$		$\alpha_0=9^0$	$\alpha_0=20^0$	$\alpha_0=30^0$
$\alpha_m=0.32715^0$	$\alpha_m=2.0283^0$	$\alpha_m=17.1840^0$		$\alpha_m=1.3187^0$	$\alpha_m=15.8944^0$	$\alpha_m=26.9403^0$
$\alpha_M=31.7180^0$	$\alpha_M=32.3202^0$	$\alpha_M=39.1803^0$		$\alpha_M=14.9648^0$	$\alpha_M=23.6366^0$	$\alpha_M=32.8262^0$
$\eta=0.9029$	$\eta=0.8938$	$\eta=0.7702$		$\eta=0.9754$	$\eta=0.8845$	$\eta=0.7513$

6. CONCLUSIONS

The efficiency (of the gear) increases when the number of teeth for the driving wheel 1, z_1, increases too and when the pressure angle, α_0, diminishes; z_2 or i_{12} are not so much influence about the efficiency value;

It can easily see that for the value $\alpha_0=20^0$, the efficiency takes roughly the value $\eta\approx0.89$ for any values of the others parameters (this justifies the choice of this value, $\alpha_0=20^0$, for the standard pressure angle of reference).

The better efficiency may be obtained only for a $\alpha_0\neq20^0$.

But the pressure angle of reference, α_0, can be decreased the same time the number of teeth for the driving wheel 1, z_1, increases, to increase the gears' efficiency;

Contrary, when we desire to create a gear with a low z_1 (for a less gauge), it will be necessary to increase the α_0 value, for maintaining a positive value for α_m (in this case the gear efficiency will be diminished);

When β increases, the efficiency, η, increases too, but the growth is insignificant.

The module of the gear, m, has not any influence on the gear's efficiency value.

When α_0 is diminished it can take a higher normal module, for increasing the addendum of teeth, but the increase of the module m at the same time with the increase of the z_1 can lead to a greater gauge.

The gears' efficiency, η, is really a function of α_0 and z_1: $\eta = f(\alpha_0, z_1)$; α_m and α_M are just the intermediate parameters.

For a good projection of the gear, it's necessary a z_1 and a z_2 greater than 30-60; but this condition may increase the gauge of mechanism.

In this chapter it determines precisely, the dynamics-efficiency, but at the gears transmissions, the dynamics efficiency is the same like the mechanical efficiency; this is a greater advantage of the gears transmissions. This advantage, specifically of the gear's mechanisms, may be found at the cam mechanisms with plate followers as well.

REFERENCES

[1] Pelecudi, Chr., ş.a., *Mecanisme.* E.D.P., Bucureşti, 1985.
[2] Petrescu, V., Petrescu, I., *Randamentul cuplei superioare de la angrenajele cu roţi dinţate cu axe fixe,* In: The Proceedings of 7th National Symposium PRASIC, Braşov, vol. I, pp. 333-338, 2002.
[3] Petrescu, R., Petrescu, F., *The gear synthesis with the best efficiency,* In: The Proceedings of ESFA'03, Bucharest, vol. 2, pp. 63-70, 2003.

CHAPTER II

GEAR DYNAMIC SYNTHESIS

Abstract: *In this chapter one succinctly presents an original method to obtain the efficiency of geared transmissions in function of the contact ratio of the gearing. With the presented relations it can make the dynamic synthesis of geared transmissions, having in view increasing the efficiency of gearing mechanisms.*
Keywords: *gear efficiency, gear dynamics, contact ratio, gear synthesis*

1 Introduction

In this chapter one presents shortly an original method to obtain the efficiency of the geared transmissions in function of the contact ratio. With the presented relations it can make the dynamic synthesis of the geared transmissions having in view increasing the efficiency of gearing mechanisms in work.

2 Determining of gearing efficiency in function of the contact ratio

We calculate the efficiency of a geared transmission, having in view the fact that at one moment there are several couples of teeth in contact, and not just one.

The start model has got four pairs of teeth in contact (4 couples) concomitantly.

The first couple of teeth in contact has the contact point i, defined by the ray r_{i1}, and the pressure angle α_{i1}; the forces which act at this point are: the motor force F_{mi}, perpendicular to the position vector r_{i1} at i and the force transmitted from the wheel 1 to the wheel 2 through the point i, $F_{\tau i}$, parallel to the path of action and with the sense from the wheel 1 to the wheel 2, the transmitted force being practically the projection of the motor force on the path of action; the defined velocities are similar to the forces (having in view the original kinematics, or the precise kinematics adopted); the same parameters will be defined for the next three points of contact, j, k, l (Fig. 1).

For starting we write the relations between the velocities (1):

$$v_{\tau i} = v_{mi} \cdot \cos \alpha_i = r_i \cdot \omega_1 \cdot \cos \alpha_i = r_{b1} \cdot \omega_1$$
$$v_{\tau j} = v_{mj} \cdot \cos \alpha_j = r_j \cdot \omega_1 \cdot \cos \alpha_j = r_{b1} \cdot \omega_1$$
$$v_{\tau k} = v_{mk} \cdot \cos \alpha_k = r_k \cdot \omega_1 \cdot \cos \alpha_k = r_{b1} \cdot \omega_1 \qquad (1)$$
$$v_{\tau l} = v_{ml} \cdot \cos \alpha_l = r_l \cdot \omega_1 \cdot \cos \alpha_l = r_{b1} \cdot \omega_1$$

From relations (1), one obtains the equality of the tangential velocities (2), and makes explicit the motor velocities (3):

$$v_{\tau i} = v_{\tau j} = v_{\tau k} = v_{\tau l} = r_{b1} \cdot \omega_1 \qquad (2)$$

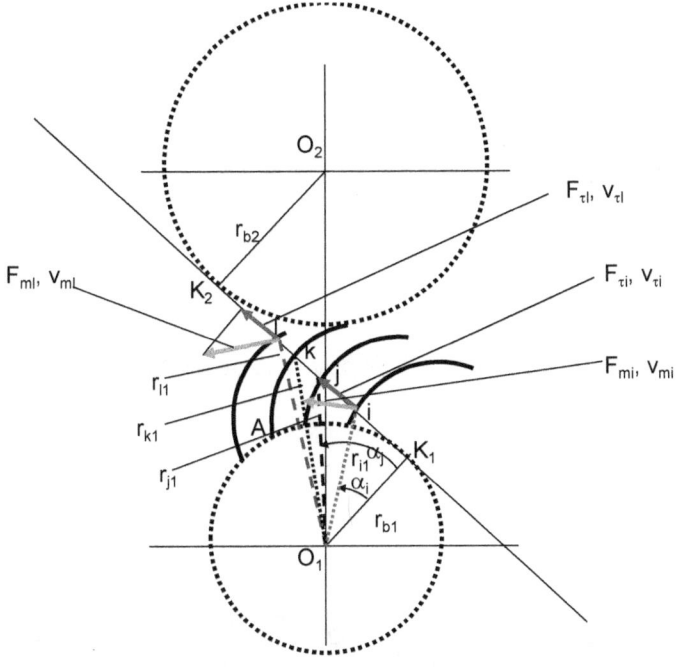

Fig. 1 Four pairs of teeth in contact concomitantly

$$v_{mi} = \frac{r_{b1} \cdot \omega_1}{\cos \alpha_i}; v_{mj} = \frac{r_{b1} \cdot \omega_1}{\cos \alpha_j}; v_{mk} = \frac{r_{b1} \cdot \omega_1}{\cos \alpha_k}; v_{ml} = \frac{r_{b1} \cdot \omega_1}{\cos \alpha_l} \qquad (3)$$

The forces transmitted concomitantly at the four points must be the same (4):

$$F_{\tau i} = F_{\tau j} = F_{\tau k} = F_{\tau l} = F_{\tau} \qquad (4)$$

The motor forces are (5):

$$F_{mi} = \frac{F_\tau}{\cos \alpha_i}; F_{mj} = \frac{F_\tau}{\cos \alpha_j}; F_{mk} = \frac{F_\tau}{\cos \alpha_k}; F_{ml} = \frac{F_\tau}{\cos \alpha_l} \qquad (5)$$

The momentary efficiency can be written in the form (6).

$$\eta_i = \frac{P_u}{P_c} = \frac{P_\tau}{P_m} = \frac{F_{\tau i} \cdot v_{\tau i} + F_{\tau j} \cdot v_{\tau j} + F_{\tau k} \cdot v_{\tau k} + F_{\tau l} \cdot v_{\tau l}}{F_{mi} \cdot v_{mi} + F_{mj} \cdot v_{mj} + F_{mk} \cdot v_{mk} + F_{ml} \cdot v_{ml}} =$$

$$= \frac{4 \cdot F_\tau \cdot r_{b1} \cdot \omega_1}{\dfrac{F_\tau \cdot r_{b1} \cdot \omega_1}{\cos^2 \alpha_i} + \dfrac{F_\tau \cdot r_{b1} \cdot \omega_1}{\cos^2 \alpha_j} + \dfrac{F_\tau \cdot r_{b1} \cdot \omega_1}{\cos^2 \alpha_k} + \dfrac{F_\tau \cdot r_{b1} \cdot \omega_1}{\cos^2 \alpha_l}} =$$

$$= \frac{4}{\dfrac{1}{\cos^2 \alpha_i} + \dfrac{1}{\cos^2 \alpha_j} + \dfrac{1}{\cos^2 \alpha_k} + \dfrac{1}{\cos^2 \alpha_l}} =$$

$$= \frac{4}{4 + tg^2 \alpha_i + tg^2 \alpha_j + tg^2 \alpha_k + tg^2 \alpha_l} \qquad (6)$$

Relations (7) and (8) are auxiliary relations:

$$K_1 i = r_{b1} \cdot tg\alpha_i; K_1 j = r_{b1} \cdot tg\alpha_j;$$

$$K_1 k = r_{b1} \cdot tg\alpha_k; K_1 l = r_{b1} \cdot tg\alpha_l$$

$$K_1 j - K_1 i = r_{b1} \cdot (tg\alpha_j - tg\alpha_i);$$

$$K_1 j - K_1 i = r_{b1} \cdot \frac{2 \cdot \pi}{z_1} \Rightarrow tg\alpha_j = tg\alpha_i + \frac{2 \cdot \pi}{z_1}$$

$$K_1 k - K_1 i = r_{b1} \cdot (tg\alpha_k - tg\alpha_i); \tag{7}$$

$$K_1 k - K_1 i = r_{b1} \cdot 2 \cdot \frac{2 \cdot \pi}{z_1} \Rightarrow tg\alpha_k = tg\alpha_i + 2 \cdot \frac{2 \cdot \pi}{z_1}$$

$$K_1 l - K_1 i = r_{b1} \cdot (tg\alpha_l - tg\alpha_i);$$

$$K_1 l - K_1 i = r_{b1} \cdot 3 \cdot \frac{2 \cdot \pi}{z_1} \Rightarrow tg\alpha_l = tg\alpha_i + 3 \cdot \frac{2 \cdot \pi}{z_1}$$

$$\begin{cases} tg\alpha_j = tg\alpha_i \pm \dfrac{2 \cdot \pi}{z_1}; \\[3mm] tg\alpha_k = tg\alpha_i \pm 2 \cdot \dfrac{2 \cdot \pi}{z_1}; \\[3mm] tg\alpha_l = tg\alpha_i \pm 3 \cdot \dfrac{2 \cdot \pi}{z_1} \end{cases} \tag{8}$$

One keeps relations (8), with the sign plus (+) for the gearing where the drive wheel 1 has external teeth (at the external or internal gearing), and with the sign (-) for the gearing where the drive wheel 1, has internal teeth (the drive wheel is a ring, only for the internal gearing).

The relation of the momentary efficiency (6) uses the auxiliary relations (8) and takes the form (9).

$$\eta_i = \frac{4}{4 + tg^2\alpha_i + tg^2\alpha_j + tg^2\alpha_k + tg^2\alpha_l} =$$

$$= \frac{4}{4 + tg^2\alpha_i + (tg\alpha_i \pm \dfrac{2\pi}{z_1})^2 + (tg\alpha_i \pm 2 \cdot \dfrac{2\pi}{z_1})^2 + (tg\alpha_i \pm 3 \cdot \dfrac{2\pi}{z_1})^2} =$$

$$= \frac{4}{4 + 4 \cdot tg^2\alpha_i + \dfrac{4\pi^2}{z_1^2} \cdot (0^2 + 1^2 + 2^2 + 3^2) \pm 2 \cdot tg\alpha_i \cdot \dfrac{2\pi}{z_1} \cdot (0 + 1 + 2 + 3)} =$$

$$= \frac{1}{1 + tg^2\alpha_i + \dfrac{4\pi^2}{E \cdot z_1^2} \cdot \displaystyle\sum_{i=1}^{E}(i-1)^2 \pm 2 \cdot tg\alpha_i \cdot \dfrac{2\pi}{E \cdot z_1} \cdot \displaystyle\sum_{i=1}^{E}(i-1)} =$$

$$= \frac{1}{1 + tg^2\alpha_1 + \dfrac{4\pi^2}{E \cdot z_1^2} \cdot \dfrac{E \cdot (E-1) \cdot (2 \cdot E - 1)}{6} \pm \dfrac{4\pi \cdot tg\alpha_1}{E \cdot z_1} \cdot \dfrac{E \cdot (E-1)}{2}} =$$

$$= \frac{1}{1 + tg^2\alpha_1 + \dfrac{2\pi^2 \cdot (E-1) \cdot (2E-1)}{3 \cdot z_1^2} \pm \dfrac{2\pi \cdot tg\alpha_1 \cdot (E-1)}{z_1}} = \qquad (9)$$

$$= \frac{1}{1 + tg^2\alpha_1 + \dfrac{2\pi^2}{3 \cdot z_1^2} \cdot (\varepsilon_{12} - 1) \cdot (2 \cdot \varepsilon_{12} - 1) \pm \dfrac{2\pi \cdot tg\alpha_1}{z_1} \cdot (\varepsilon_{12} - 1)}$$

In expression (9) one starts with relation (6) where four pairs are in contact concomitantly, but then one generalizes the expression, replacing the 4 figure (four pairs) with E couples, replacing figure 4 with the E variable, which represents the whole number of the contact ratio +1, and after restricting the sums

24

expressions, we replace the variable E with the contact ratio ε_{12}, as well.

The mechanical efficiency offers more advantages than the momentary efficiency, and will be calculated approximately, by replacing in relation (9) the pressure angle α_1, with the normal pressure angle α_0 the relation taking the form (10); where ε_{12} represents the contact ratio of the gearing, and it will be calculated with expression (11) for the external gearing, and with relation (12) for the internal gearing.

$$\eta_m = \frac{1}{1 + tg^2\alpha_0 + \dfrac{2\pi^2}{3 \cdot z_1^2} \cdot (\varepsilon_{12} - 1) \cdot (2 \cdot \varepsilon_{12} - 1) \pm \dfrac{2\pi \cdot tg\alpha_0}{z_1} \cdot (\varepsilon_{12} - 1)} \quad (10)$$

$$\varepsilon_{12}^{a.e.} = \frac{\sqrt{z_1^2 \cdot \sin^2\alpha_0 + 4 \cdot z_1 + 4} + \sqrt{z_2^2 \cdot \sin^2\alpha_0 + 4 \cdot z_2 + 4} - (z_1 + z_2) \cdot \sin\alpha_0}{2 \cdot \pi \cdot \cos\alpha_0} \quad (11)$$

$$\varepsilon_{12}^{a.i.} = \frac{\sqrt{z_e^2 \cdot \sin^2\alpha_0 + 4 \cdot z_e + 4} - \sqrt{z_i^2 \cdot \sin^2\alpha_0 - 4 \cdot z_i + 4} + (z_i - z_e) \cdot \sin\alpha_0}{2 \cdot \pi \cdot \cos\alpha_0} \quad (12)$$

The calculations made have been centralized in the table 1.

Table 1								
The centralized results								
z1	α_0	z2	ε_{12}^{ae}	η_{12}^{ae}	η_{21}^{ae}	ε_{12}^{ai}	η_{12}^{ai}	η_{21}^{ai}
42	20	126	1.79	0.844	0.871	1.92	0.838	0.895
46	19	138	1.87	0.856	0.882	2.00	0.850	0.905
52	18	156	1.96	0.869	0.893	2.09	0.864	0.915
58	17	174	2.06	0.880	0.904	2.20	0.876	0.925
65	16	195	2.17	0.892	0.914	2.32	0.887	0.933
74	15	222	2.30	0.903	0.923	2.46	0.899	0.942
85	14	255	2.44	0.914	0.933	2.62	0.910	0.949
98	13	294	2.62	0.924	0.941	2.81	0.920	0.956
115	12	345	2.82	0.934	0.949	3.02	0.931	0.963
137	11	411	3.06	0.943	0.957	3.28	0.941	0.969
165	10	495	3.35	0.952	0.964	3.59	0.950	0.974
204	9	510	3.68	0.960	0.970	4.02	0.958	0.980
257	8	514	4.09	0.968	0.975	4.57	0.966	0.985
336	7	672	4.66	0.975	0.980	5.21	0.973	0.989
457	6	914	5.42	0.981	0.985	6.06	0.980	0.992
657	5	1314	6.49	0.986	0.989	7.26	0.986	0.994

3 Determining of gearing efficiency in function of the contact ratio to the bended teeth

Generally we use gearings with teeth inclined (with bended teeth). For gears with bended teeth, the calculations show a decrease in yield when the inclination angle increases. For angles with inclination which not exceed 25 degree the efficiency of gearing is good (see the table 2). When the inclination angle (β) exceeds 25 degrees the gearing will suffer a significant drop in yield (see the tables 3-4).

Table 2. *Bended teeth, β=25 [deg].*

z1	α_0 [grad]	z2	ε_{12}^{ae}	η_{12}^{ae}	η_{21}^{ae}	ε_{12}^{ai}	η_{12}^{ai}	η_{21}^{ai}
\multicolumn{9}{}{Determining the efficiency when $\beta=25$ [deg]}								
42	20	126	1,708	0,829	0,851	1,791	0,826	0,871
46	19	138	1,776	0,843	0,864	1,865	0,839	0,883
52	18	156	1,859	0,856	0,876	1,949	0,853	0,895
58	17	174	1,946	0,869	0,889	2,043	0,866	0,906
65	16	195	2,058	0,882	0,900	2,151	0,879	0,917
74	15	222	2,165	0,894	0,911	2,275	0,892	0,927
85	14	255	2,299	0,906	0,922	2,418	0,904	0,936
98	13	294	2,456	0,917	0,932	2,584	0,915	0,945
115	12	345	2,641	0,928	0,941	2,780	0,926	0,953
137	11	411	2,863	0,938	0,950	3,013	0,937	0,961
165	10	495	3,129	0,948	0,958	3,295	0,947	0,968
204	9	510	3,443	0,957	0,965	3,665	0,956	0,974
257	8	514	3,829	0,965	0,971	4,146	0,964	0,981
336	7	672	4,357	0,973	0,977	4,719	0,972	0,985
457	6	914	5,064	0,980	0,983	5,486	0,979	0,989
657	5	1314	6,056	0,985	0,988	6,563	0,985	0,992

Table 3. *Bended teeth, β=35 [deg].*

z1	α_0 [grad]	z2	ε_{12}^{ae}	η_{12}^{ae}	η_{21}^{ae}	ε_{12}^{ai}	η_{12}^{ai}	η_{21}^{ai}
		Determining the efficiency when $\beta=35$ [deg]						
42	20	126	1,620	0,809	0,827	1,677	0,807	0,843
46	19	138	1,681	0,825	0,841	1,741	0,822	0,858
52	18	156	1,755	0,840	0,856	1,815	0,838	0,871
58	17	174	1,832	0,854	0,870	1,898	0,852	0,885
65	16	195	1,948	0,868	0,883	1,993	0,867	0,897
74	15	222	2,030	0,882	0,896	2,103	0,881	0,909
85	14	255	2,150	0,895	0,909	2,230	0,894	0,921
98	13	294	2,293	0,908	0,920	2,379	0,907	0,932
115	12	345	2,461	0,920	0,931	2,554	0,919	0,942
137	11	411	2,663	0,932	0,942	2,764	0,931	0,951
165	10	495	2,906	0,942	0,951	3,017	0,942	0,959
204	9	510	3,196	0,952	0,959	3,345	0,952	0,968
257	8	514	3,556	0,962	0,967	3,766	0,961	0,975
336	7	672	4,041	0,970	0,974	4,281	0,969	0,981
457	6	914	4,692	0,978	0,981	4,971	0,977	0,986
657	5	1314	5,607	0,984	0,986	5,942	0,984	0,990

Table 4. *Bended teeth, β=45 [deg].*

Determining the efficiency when $\beta=45$ [deg]								
$z1$	α_0 [grad]	$z2$	ε_{12}^{ae}	η_{12}^{ae}	η_{21}^{ae}	ε_{12}^{ai}	η_{12}^{ai}	η_{21}^{ai}
42	20	126	1,505	0,772	0,784	1,539	0,771	0,796
46	19	138	1,555	0,790	0,802	1,590	0,789	0,814
52	18	156	1,618	0,808	0,820	1,650	0,807	0,831
58	17	174	1,680	0,825	0,837	1,718	0,824	0,848
65	16	195	1,810	0,841	0,853	1,796	0,841	0,864
74	15	222	1,848	0,858	0,869	1,888	0,858	0,879
85	14	255	1,949	0,874	0,884	1,994	0,874	0,894
98	13	294	2,070	0,889	0,899	2,119	0,889	0,908
115	12	345	2,215	0,904	0,913	2,268	0,903	0,921
137	11	411	2,389	0,918	0,926	2,446	0,917	0,933
165	10	495	2,600	0,931	0,938	2,662	0,930	0,944
204	9	510	2,855	0,943	0,948	2,938	0,943	0,955
257	8	514	3,173	0,954	0,958	3,290	0,954	0,965
336	7	672	3,599	0,964	0,967	3,732	0,964	0,973
457	6	914	4,171	0,973	0,976	4,325	0,973	0,980
657	5	1314	4,976	0,981	0,983	5,161	0,981	0,986

New calculation relationships can be put in the forms (13-15).

$$\eta_m = \frac{z_1^2 \cdot \cos^2 \beta}{z_1^2 (tg^2 \alpha_0 + \cos^2 \beta) + \frac{2}{3}\pi^2 \cos^4 \beta (\varepsilon - 1)(2\varepsilon - 1) \pm 2\pi tg \alpha_0 z_1 \cos^2 \beta (\varepsilon - 1)} \qquad (13)$$

$$\varepsilon^{a.e.} = \frac{1 + tg^2 \beta}{2 \cdot \pi} \cdot$$
$$\cdot \left\{ \sqrt{\left[(z_1 + 2 \cdot \cos \beta) \cdot tg \alpha_0 \right]^2 + 4 \cdot \cos^3 \beta \cdot (z_1 + \cos \beta)} + \right.$$
$$+ \sqrt{\left[(z_2 + 2 \cdot \cos \beta) \cdot tg \alpha_0 \right]^2 + 4 \cdot \cos^3 \beta \cdot (z_2 + \cos \beta)} -$$
$$\left. - (z_1 + z_2) \cdot tg \alpha_0 \right\} \qquad (14)$$

$$\varepsilon^{a.i.} = \frac{1 + tg^2 \beta}{2 \cdot \pi} \cdot$$
$$\cdot \left\{ \sqrt{\left[(z_e + 2 \cdot \cos \beta) \cdot tg \alpha_0 \right]^2 + 4 \cdot \cos^3 \beta \cdot (z_e + \cos \beta)} - \right.$$
$$- \sqrt{\left[(z_i - 2 \cdot \cos \beta) \cdot tg \alpha_0 \right]^2 - 4 \cdot \cos^3 \beta \cdot (z_i - \cos \beta)} -$$
$$\left. - (z_e - z_i) \cdot tg \alpha_0 \right\} \qquad (15)$$

The calculation relationships (13-15) are general. They have the advantage that can be used with great precision in determining the efficiency of any type of gearings.

To use them at the gearing without bended teeth is enough to assign them a beta value = zero. The results obtained in this case will be identical to the ones of the relations 10-12.

4 Conclusions

The best efficiency can be obtained with the internal gearing when the drive wheel 1 is the ring; the minimum efficiency will be obtained when the drive wheel 1 of the internal gearing has external teeth.

For the external gearing, the best efficiency is obtained when the bigger wheel is the drive wheel; *when we decrease the normal angle α_0, the contact ratio increases and the efficiency increases as well.*

The efficiency increases too, when the number of teeth of the drive wheel 1 increases (when z_1 increases).

References

[1] PETRESCU R.V., PETRESCU F.I., POPESCU N., „Determining Gear Efficiency", In Gear Solutions magazine, USA, pp. 19-28, March 2007.

CHAPTER III

PRESENTING A
"DYNAMIC ORIGINAL MODEL"
USED TO STUDY TOOTHED
GEARING WITH PARALLEL AXES

Abstract: *Nearly all the models studied the dynamic on gearing with axes parallel, is based on mechanical models of classical (known) who is studying spinning vibration of shafts gears and determine their own beats and strains of shafts spinning; sure that they are very useful, but are not actually join formed of the two teeth in contact (or more pairs of teeth in contact), that is not treated physiology of the mechanism itself with toothed gears for a view that the phenomena are dynamic taking place in top gear flat; model [1] just try this so, but the whole theory is based directly on the impact of teeth (collisions between teeth); this chapter will present an original model that tries to explore the dynamic phenomena taking place in the plane geared couple from the geared transmissions with parallel axes.*

Keywords: *Gear dynamics, spinning vibration, shaft vibration, impact of teeth, geared couple, angle characteristic*

1. Introduction (or the starting idea)

Nearly all the models [1, 2, 3, 7] studied the dynamic on gearing with axes parallel, is based on mechanical models of classical (known) who is studying spinning vibration of shafts gears and determine their own beats and strains of shafts spinning; sure that they are very useful, but are not actually join formed of the two teeth in contact (or more pairs of teeth in contact), that is not treated physiology of the mechanism itself with toothed gears for a view that the phenomena are dynamic taking place in top gear flat; model [1] just try this so, but the whole theory is based directly on the impact of teeth (collisions between teeth); this chapter will present an original model that tries to explore the dynamic phenomena taking place in the plane geared couple from the geared transmissions with parallel axes.

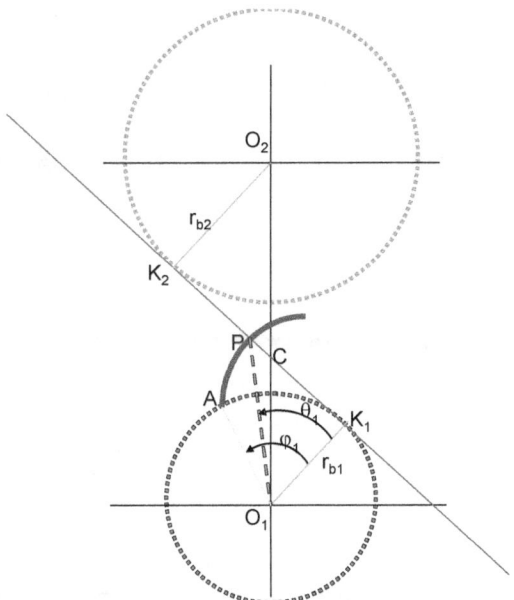

Fig. 1. *Angles characteristic at a position of a tooth from the driving wheel, in gearing*

Figure 1 presents a tooth of the lead wheel 1, in gearing, at a certain position on the gearing segment K_1K_2. It is characterized by angles θ_1 and φ_1, the first showing the position of the vector O_1P (the contact vector) in relation to fixed vector O_1K_1, and the second showing how much is turned the tooth (leading wheel 1) in relation to O_1K_1.

Between the two angles are the relations of liaison 1:

$$\varphi_1 = tg\,\theta_1 \qquad\qquad \theta_1 = arctg\,\varphi_1 \tag{1}$$

Since φ_1 is the sum of angles θ_1 and γ_1, where the angle γ_1 represents the known function $inv\theta_1$:

$$\varphi_1 = \theta_1 + \gamma_1 = \theta_1 + inv\,\theta_1 = \theta_1 + (tg\,\theta_1 - \theta_1) = tg\,\theta_1 \tag{2}$$

One derives the relations (1) and one obtains the forms (3):

$$
\left\{
\begin{aligned}
&\dot{\varphi}_1 = (1 + tg^2\theta_1)\cdot\dot{\theta}_1 = (1 + \varphi_1^2)\cdot\dot{\theta}_1 \\[2mm]
&\dot{\theta}_1 = \frac{\dot{\varphi}_1}{1 + \varphi_1^2} = D_1\cdot\omega_1 ; D_1 = \frac{1}{1 + \varphi_1^2} = \frac{1}{1 + tg^2\theta_1} \\[2mm]
&\ddot{\theta}_1 = \dot{D}_1\cdot\omega_1 = D_1'\cdot\omega_1^2 ; D_1' = \frac{-2\cdot\varphi_1}{(1 + \varphi_1^2)^2} = \frac{-2\cdot tg\,\theta_1}{(1 + tg^2\theta_1)^2}
\end{aligned}
\right. \tag{3}
$$

2. The dynamic model

The dynamic model considered [3, 4] (fig. 2) is similar to that of cam gears [4, 5, 6], as geared wheels are similar to those with camshaft; practically the toothed wheel is a multiple cam, each tooth is a lobe (cam) showing only the up lifting phase. Forces and J * (M *) is amended, so equation of motion will get another look.

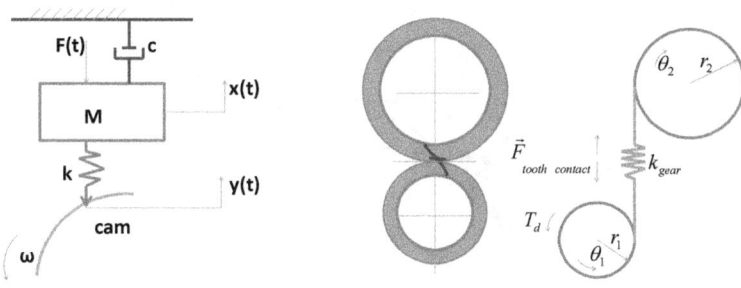

Fig. 2. *The dynamic model. Forces, displacements, and elasticity of the system*

Contact between the two teeth is practically a contact between a rotation cam and a rocker follower. So similar to models with cams [5, 6] it will determine precision cinematic (dynamics cinematic) to join with gears with parallel axes. Vector which should be the leading at the driving wheel 1 (in the dynamics cinematic), is the contact vector O_1P, the angle of his position as θ_1 and his angular velocity, $\dot{\theta}_1$. To the driven wheel 2, one forwards the speed v_2 (see schedule kinematics in Figure 3).

$$v_2 = -v_1 \cdot \cos\theta_1 = -r_{p1} \cdot \dot{\theta}_1 \cdot \cos\theta_1 = -r_{b1} \cdot \dot{\theta}_1 = -r_{b1} \cdot D_1 \cdot \omega_1$$

$$but: \quad v_2 = r_{b2} \cdot \omega_2 => \omega_2 = -\frac{r_{b1}}{r_{b2}} \cdot D_1 \cdot \omega_1$$

(4)

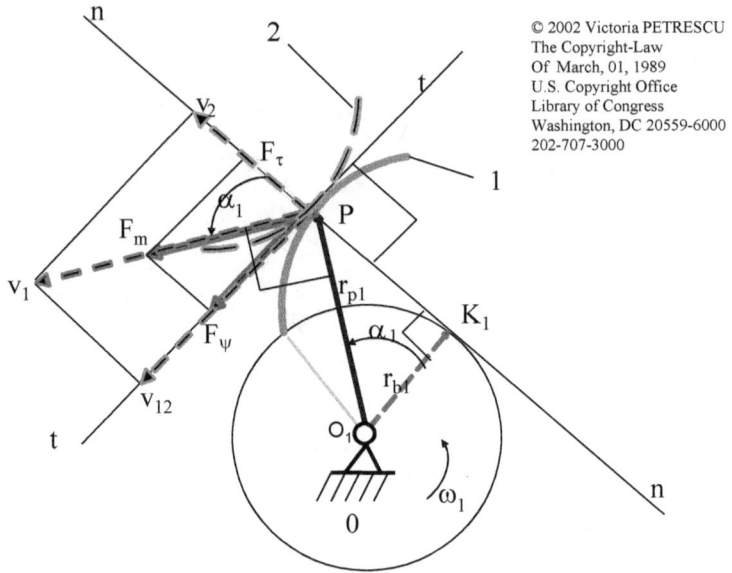

Fig. 3. *Forces and velocities characteristic at a position of a tooth from the driving wheel, in gearing*

By derivation is calculated and angular acceleration (precision acceleration), to the wheel 2 (5), and by integration, movement of the wheel 2 (6):

$$\varepsilon_2 = -\frac{r_{b1}}{r_{b2}} \cdot D_1' \cdot \omega_1^2 \tag{5}$$

$$\varphi_2 = -\frac{r_{b1}}{r_{b2}} \cdot arctg(\varphi_1) = -\frac{r_{b1}}{r_{b2}} \cdot \theta_1 \tag{6}$$

Reduced Force (engines and resistance) at the wheel 1, lead, is equal to the elastic force of couple (while at the wheel led 2 does not intervene and a strong technological force or one additional) and is written in the form (7):

$$F^* = K \cdot (r_{b1} \cdot \varphi_1 - r_{b2} \cdot \varphi_2) =$$
$$= K \cdot (r_{b1} \cdot \varphi_1 - r_{b2} \cdot \frac{r_{b1}}{r_{b2}} \cdot \theta_1) = \qquad (7)$$
$$= K \cdot r_{b1} \cdot (tg\theta_1 - \theta_1)$$

Minus sign was already taken, so φ_2 is replaced just in module in the expression of 7, and K means constant elastic of teeth in contact, and is measured in [N/m]. Dynamic equation of motion is written:

$$M^* \cdot \ddot{x} + \frac{1}{2} \cdot \frac{dM^*}{dt} \cdot \dot{x} = F^* \qquad (8)$$

Reduced mass, M^*, is determined by the relationship (9):

$$M^* = (J_1 + \frac{1}{i^2} \cdot J_2) \cdot \frac{1}{r_{p1}^2} = (J_1 + \frac{1}{i^2} \cdot J_2) \cdot \frac{\cos^2 \theta_1}{r_{b1}^2} =$$

$$= \frac{J_1 + \frac{1}{i^2} \cdot J_2}{r_{b1}^2} \cdot \cos^2 \theta_1 = C_M \cdot \cos^2 \theta_1; C_M = \frac{J_1 + \frac{1}{i^2} \cdot J_2}{r_{b1}^2} \qquad (9)$$

Where, J_1 and J_2 represent the moments of inertia (mass, mechanical), reduced to wheel 1, while i is the module of transmission ratio from the wheel 1 to the wheel 2 (see relation 10):

$$i = \frac{r_{b2}}{r_{b1}} = -\frac{\omega_1}{\omega_2} \qquad (10)$$

The moving x of the wheel 2 on the gearing segment, writes:

$$x = r_{b2} \cdot \varphi_2 = r_{b2} \cdot -\frac{r_{b1}}{r_{b2}} \cdot arctg\,\varphi_1 = -r_{b1} \cdot arctg\,\varphi_1 = -r_{b1} \cdot \theta_1 \qquad (11)$$

The speed and corresponding accelerations can be written:

$$\dot{x} = -r_{b1} \cdot \dot{\theta}_1 = -r_{b1} \cdot \frac{1}{1 + tg^2\theta_1} \cdot \omega_1 \qquad (12)$$

$$\ddot{x} = -r_{b1} \cdot \ddot{\theta}_1 = -r_{b1} \cdot \frac{-2 \cdot tg\,\theta_1}{(1 + tg^2\theta_1)^2} \cdot \omega_1^2 = 2 \cdot r_{b1} \cdot \frac{tg\,\theta_1}{(1 + tg^2\theta_1)^2} \cdot \omega_1^2 \qquad (13)$$

One derives the reduced mass and result the expression (14):

$$\frac{dM^*}{dt} = -2 \cdot C_M \cdot \frac{\cos\theta_1 \cdot \sin\theta_1}{1 + tg^2\theta_1} \cdot \omega_1 \qquad (14)$$

The equation of motion (8) takes now the form (15), which can arrange and form (16):

$$3 \cdot C_M \cdot r_{b1} \cdot \frac{tg\,\theta_1}{(1+tg^2\theta_1)^3} \cdot \omega_1^2 = K \cdot r_{b1} \cdot (tg\,\theta_1 - \theta_1) \qquad (15)$$

$$\theta_1^d \equiv \theta_1 = tg\,\theta_1 - \frac{3 \cdot C_M \cdot tg\,\theta_1 \cdot \omega_1^2}{K \cdot (1+tg^2\theta_1)^3} =$$

$$= tg\,\theta_1 \cdot [1 - \frac{3 \cdot (J_1 + \frac{r_{b1}^2}{r_{b2}^2} \cdot J_2) \cdot \omega_1^2}{r_{b1}^2 \cdot K \cdot (1+tg^2\theta_1)^3}] \qquad (16)$$

The expression (16) is the solution equation of motion of superior couple; to an angle of rotation of the wheel 1, φ_1, known, which is corresponding a pressure angle θ_1 known, the expression (16) generates a dynamic angle of pressure, θ_1^d.

In terms of the constant elasticity of the teeth in contact, K, is large enough, if the radius of the base circle of the wheel 1 don't decrease too much (z_1 to be greater than 15-20), for normal speeds and even higher (but not too large), the ratio of expression parenthesis 16 remains under the value 1, and even much lower than 1, and the expression 16 can be engineering estimated to the natural short form (17):

$$\theta_1^d = tg\,\theta_1 = \varphi_1^c \equiv \varphi_1 \qquad (17)$$

3. The (dynamic) angular velocity at the lead wheel 1

We can determine now the instantaneous (momentary) angular velocity of the lead wheel 1 (relationship 19); it used the intermediate relation (18) as well:

$$\frac{\Delta\omega_1}{\omega_m} = \frac{\Delta\varphi_1}{\varphi_1} \Rightarrow \Delta\omega_1 = \frac{\Delta\varphi_1}{\varphi_1}\cdot\omega_m = \frac{\varphi_1^d - \varphi_1}{\varphi_1}\cdot\omega_m =$$
$$= \frac{tg(\theta_1^d) - \varphi_1}{\varphi_1}\cdot\omega_m = \frac{tg(\varphi_1) - \varphi_1}{\varphi_1}\cdot\omega_m = \frac{inv\,\varphi_1}{\varphi_1}\cdot\omega_m \qquad (18)$$

$$\omega_1 = \omega_m + \Delta\omega_1 = (1 + \frac{inv\,\varphi_1}{\varphi_1})\cdot\omega_m =$$
$$= \frac{tg(\varphi_1)}{\varphi_1}\cdot\omega_m = \frac{tg(tg\,\theta_1)}{tg\,\theta_1}\cdot\omega_m = R_{d1}\cdot\omega_m \qquad (19)$$

It defines the dynamic coefficient, R_{d1}, as the ratio between the tangent of the angle φ_1 and φ_1 angle, or the ratio, $\dfrac{tg(tg\,\theta_1)}{tg\,\theta_1}$, relationship 1.20:

$$R_{d1} = \frac{tg(tg\theta_1)}{tg\theta_1}$$ (20)

Dynamic synthesis of gearing with axes parallel can be made taking into account the relation (1.20). The necessity of obtaining a dynamic factor as low (close to the value 1), requires limiting the maximum pressure angle, θ_{1M} and the normal angle α_0, and increasing the minimum number of teeth of the leading wheel, 1, z_{1min}.

4. The dynamic of wheel 2 (conducted)

One can determine now the instantaneous (momentary) dynamic angular velocity at the led wheel 2 (relationship 28), and all angular parameters (displacement, velocity, acceleration), in three situations: classical cinematic, precision cinematic, and dynamic; where: c=cinematic, cp=precision cinematic, d=dynamic (see the relation: 21-31).

$$\varphi_2^c = -\frac{r_{b1}}{r_{b2}} \cdot \varphi_1$$ (21)

$$\omega_2^c = -\frac{r_{b1}}{r_{b2}} \cdot \omega_1$$ (22)

$$\varepsilon_2^c = -\frac{r_{b1}}{r_{b2}} \cdot \varepsilon_1 = 0$$ (23)

$$\varphi_2^{cp} = -\frac{r_{b1}}{r_{b2}} \cdot arctg\,\varphi_1 = -\frac{r_{b1}}{r_{b2}} \cdot \theta_1 \qquad (24)$$

$$\omega_2^{cp} = -\frac{r_{b1}}{r_{b2}} \cdot \frac{1}{1+\varphi_1^2} \cdot \omega_1 = -\frac{r_{b1}}{r_{b2}} \cdot \frac{1}{1+tg^2\theta_1} \cdot \omega_1 \qquad (25)$$

$$\varepsilon_2^{cp} = -\frac{r_{b1}}{r_{b2}} \cdot \frac{-2 \cdot \varphi_1}{(1+\varphi_1^2)^2} \cdot \omega_1^2 = -\frac{r_{b1}}{r_{b2}} \cdot \frac{-2 \cdot tg\,\theta_1}{(1+tg^2\theta_1)^2} \cdot \omega_1^2 \qquad (26)$$

Dynamics of wheel 2 (conducted) are calculated with relations (27-31):

$$\varphi_2^d = -\frac{r_{b1}}{r_{b2}} \cdot \int \frac{tg\,\varphi_1}{\varphi_1 + \varphi_1^3} d\varphi_1 \qquad (27)$$

$$\omega_2^d = -\frac{r_{b1}}{r_{b2}} \cdot \frac{1}{1+\varphi_1^2} \cdot \frac{tg\,\varphi_1}{\varphi_1} \cdot \omega_1 = -\frac{r_{b1}}{r_{b2}} \cdot \frac{1}{1+tg^2\theta_1} \cdot \frac{tg(tg\,\theta_1)}{tg\,\theta_1} \cdot \omega_1 \qquad (28)$$

$$\varepsilon_2^d = -\frac{r_{b1}}{r_{b2}} \cdot \frac{(1+tg^2\varphi_1) \cdot (\varphi_1 + \varphi_1^3) - tg\,\varphi_1 \cdot (1+3 \cdot \varphi_1^2)}{(\varphi_1 + \varphi_1^3)^2} \cdot \omega_1^2 \qquad (29)$$

With:

$$\varphi_{1m} = tg\,\theta_{1m} = \frac{(z_1 + z_2) \cdot \sin\alpha_0 - \sqrt{z_2^2 \cdot \sin^2\alpha_0 + 4 \cdot z_2 + 4}}{z_1 \cdot \cos\alpha_0} \qquad (30)$$

42

$$\varphi_{1M} = tg\,\theta_{1M} = \frac{\sqrt{z_1^2 \cdot \sin^2 \alpha_0 + 4 \cdot z_1 + 4}}{z_1 \cdot \cos \alpha_0} \qquad (31)$$

Can be defined to wheel 2 a dynamic coefficient Rd2, (see the relations 28 and 32):

$$R_{d2} = \frac{1}{1+\varphi_1^2} \cdot \frac{tg\,\varphi_1}{\varphi_1} = \frac{1}{1+tg^2\theta_1} \cdot \frac{tg\,(tg\,\theta_1)}{tg\,\theta_1} \qquad (32)$$

5. Conclusions

Representation of angular velocity, ω_2, depending on the angle φ_1, for r_{b1} and r_{b2} known (z_1, z_2, m, and α_0 imposed), and for a certain amount of angular velocity input constant (imposed by the speed of the shaft which is mounted wheel leading 1), can be seen in the figures 4: a, b.

Observe the appearance of vibration at the dynamic angular velocity, ω_2.

Start with equal rays and 20 degrees all for, α_0 (fig. 4a), and stay on the chart last rays equal and α_0 reduced to 5 degrees (fig. 4b).

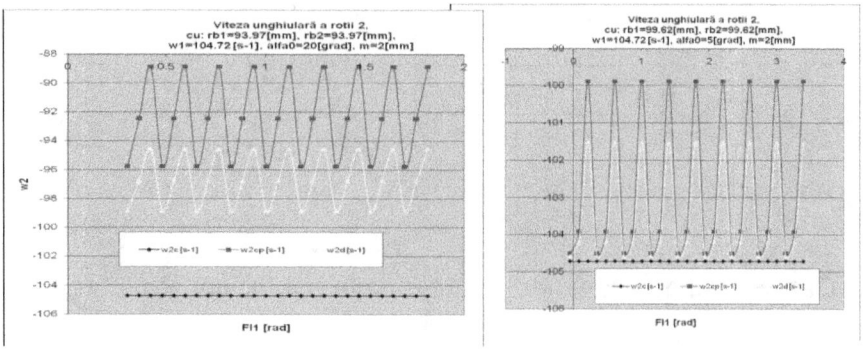

Fig. 4. *Dynamics of wheel 2; ω_2 cinematic, ω_2 in precision cinematic, ω_2 dynamic*

In fig. 5 it can see the experimental vibration [1, 2].

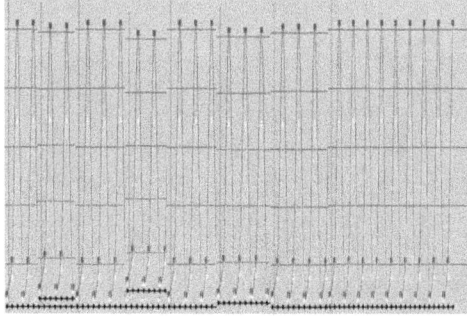

Fig. 5. *Dynamics of wheel 2; ω_2 dynamic (the vibrations obtained experimental)*

The presented (original) method is more simply, directly, naturaly and rapidly than the classics.

References

[1]-Bajer, A., *„Parallel finite element simulator of planetary gear trains",* In Ph. D. Dissertation, The University of Texas, 2001;

[2]-Li, J., *„Gear Fatigue Crack Prognosis Using Embedded Model, Gear Dynamic Model and Fracture Mechanics",* Department of Mechanical, Aerospace and Nuclear Engineering, Rensselaer Polytechnic Institute;

[3]-Peeters, J., Vandepitte, P., *„Flexible multibody model of a three-stage planetary gear-box in a wind turbine",* In Proceedings of ISMA, 2004, p. 3923-3942;

[4]-PETRESCU, F.I., PETRESCU, R.V. *Contributions at the dynamics of cams.* In the Ninth IFToMM International Symposium on Theory of Machines and Mechanisms, SYROM 2005, Bucharest, Romania, 2005, Vol. I, p. 123-128;

[5]-PETRESCU, R.V., COMĂNESCU A., PETRESCU F.I., *Dynamics of Cam Gears Illustrated at the Classic Distribution Mechanism.* In NEW TRENDS IN MECHANISMS, Ed. Academica - Greifswald, 2008, I.S.B.N. 978-3-9402-37-10-1;

[6]-PETRESCU, R.V., COMĂNESCU A., PETRESCU F.I., ANTONESCU O., *The Dynamics of Cam Gears at the Module B (with Translated Follower with Roll).* In NEW TRENDS IN MECHANISMS, Ed. Academica - Greifswald, 2008, I.S.B.N. 978-3-9402-37-10-1.

[7]-Takeuchi, T., Togai, K., *„Gear Whine Analysis with Virtual Power Train",* In Mitsubishi Motors Technical Review, 2004, No. 16, p. 23-28.

[8]-PETRESCU F.I., PETRESCU R.V., *Presenting A Dynamic Original Model Used to Study Toothed Gearing with Parallel Axes.* In the 13-th edition of SCIENTIFIC CONFERENCE with international participation, Constantin Brâncusi University, Târgu-Jiu, November 2008.

CHAPTER IV

PLANETARY TRAINS EFFICIENCY

Abstract: Synthesis of classical planetary mechanisms is usually based on kinematic relations, considering in especially the transmission ratio input-output achieved. The planetary mechanisms are less synthesized based by their mechanical efficiency developed in operation, although this criterion is part of the real dynamics of mechanisms, being also the most important criterion in terms of performance of a mechanism. Even when it used the efficiency criterion, the determination of the planetary yield, is made only with approximate relationships. The most widely recognized method is one method of Russian school of mechanisms. This chapter determines the exact method to calculate the mechanical efficiency of a planetary mechanism. In this mode it is resolving one important problem of the dynamics of planetary mechanisms.

Keywords: planetary efficiency, planetary kinematic, planetary synthesis

1. Introduction

Synthesis of classical planetary mechanisms is usually based on kinematic relations, considering in especially the transmission ratio input-output achieved. The most common model used is the differential planetary mechanism showed in Figure 1.

Usually the formula 1 is determined by writing the relationship Willis (1').

For the various cinematic planetary systems presented in Figure 3, where entry is made by the planetary carrier (H), and output is achieved by the final element (f), the initial element being usually immobilized, will be used for the kinematic calculations the relationships generalized 1 and 2.

The planetary mechanisms are less synthesized based by their mechanical efficiency developed in operation, although this criterion is part of the real dynamics of mechanisms, being also the most important criterion in terms of performance of a mechanism.

Even when it used the efficiency criterion, the determination of the planetary yield, is made only with approximate relationships. The most widely recognized method is one method of Russian school of mechanisms. This chapter determines the real efficiency of the planetary trains (determines the exact method).

2. Kinematic synthesis

Synthesis of classical planetary mechanisms is usually based on kinematic relations, considering in especially the transmission ratio input-output achieved. The most common model used is the differential planetary mechanism showed in the Figure 1.

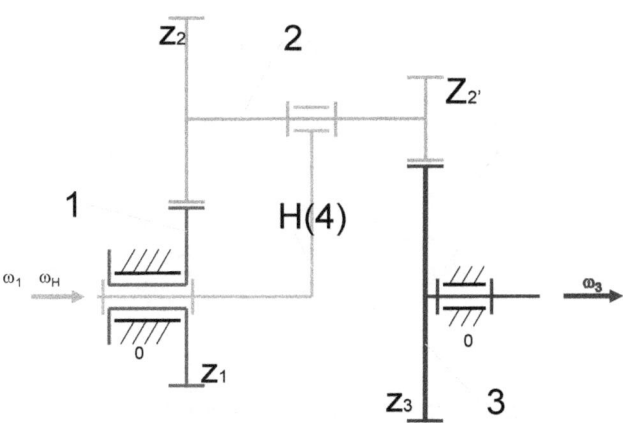

Fig. 1. *Kinematic schema of a differential planetary mechanism (M=2)*

For this mechanism to have one single degree of mobility, remaining in use with a drive desmodromic unique and single output, it is necessary to reduce the mobility of the mechanism from two to one, which can be obtained by connecting in series or parallel of two or more planetary gear, by binding to gears with fixed axes, or the hardening of one of its mobile elements; element 1 in this case (case in which the wheel 1 is identified with the fixed element 0; fig. 2).

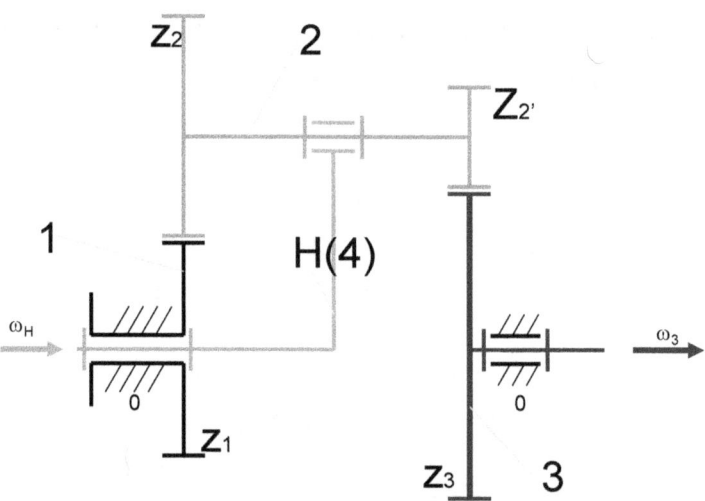

Fig. 2. *Kinematic schema of a simple planetary mechanism (M=1)*

The entrance to the simple planetary from Figure 2 is made by the planetary carrier (H), and the output is done by the mobile cinematic element (3), the wheel (3). Kinematic ratio between input-output (H-3) is written directly with the relationship 1.

$$i_{H3}^1 = \frac{1}{i_{3H}^1} = \frac{1}{1 - i_{31}^H} = \frac{1}{1 - \dfrac{1}{i_{13}^H}}$$ (1)

Where i_{13}^H is the ratio of transmission input output corresponding to the mechanism with fixed axis (when the planetary carrier H is fixed), and is determined in function of the cinematic schema of planetary gear used; for the model in Figure 2 it is determined by the relation 2, as a depending on the numbers of teeth of the wheels 1, 2, 2 ', 3.

$$i_{13}^H = \frac{z_2}{z_1} \cdot \frac{z_3}{z_{2'}}$$ (2)

Usually the formula 1 is determined by writing the relationship Willis (1').

$$\begin{cases} i^H_{13} = \dfrac{\omega_1 - \omega_H}{\omega_3 - \omega_H} \equiv \dfrac{z_2}{z_1} \cdot \dfrac{z_3}{z_{2'}} \\[4mm] \dfrac{z_2}{z_1} \cdot \dfrac{z_3}{z_{2'}} = \dfrac{\dfrac{\omega_1}{\omega_H} - \dfrac{\omega_H}{\omega_H}}{\dfrac{\omega_3}{\omega_H} - \dfrac{\omega_H}{\omega_H}} \\[8mm] i^H_{13} = \dfrac{z_2 \cdot z_3}{z_1 \cdot z_{2'}} = \dfrac{0-1}{\dfrac{\omega_3}{\omega_H} - 1} = \dfrac{1}{1 - i_{3H}} = \dfrac{1}{1 - \dfrac{1}{i^1_{H3}}} \Rightarrow \\[8mm] \Rightarrow i^1_{H3} = \dfrac{1}{1 - \dfrac{1}{i^H_{13}}} \end{cases}$$

(1')

For the various cinematic planetary systems presented in Figure 3, where entry is made by the planetary carrier (H), and output is achieved by the final element (f), the initial element being usually immobilized, will be used for the kinematic calculations the relationships generalized 1 and 2; the relationship 1 takes the general form 3, and the relation 2 is written in one of the forms 4 particularized for each schema separately, used; where i become 1, and f takes the value 3 or 4 as appropriate.

$$i^i_{Hf} = \frac{1}{i^i_{fH}} = \frac{1}{1 - i^H_{fi}} = \frac{1}{1 - \dfrac{1}{i^H_{if}}}$$

(3)

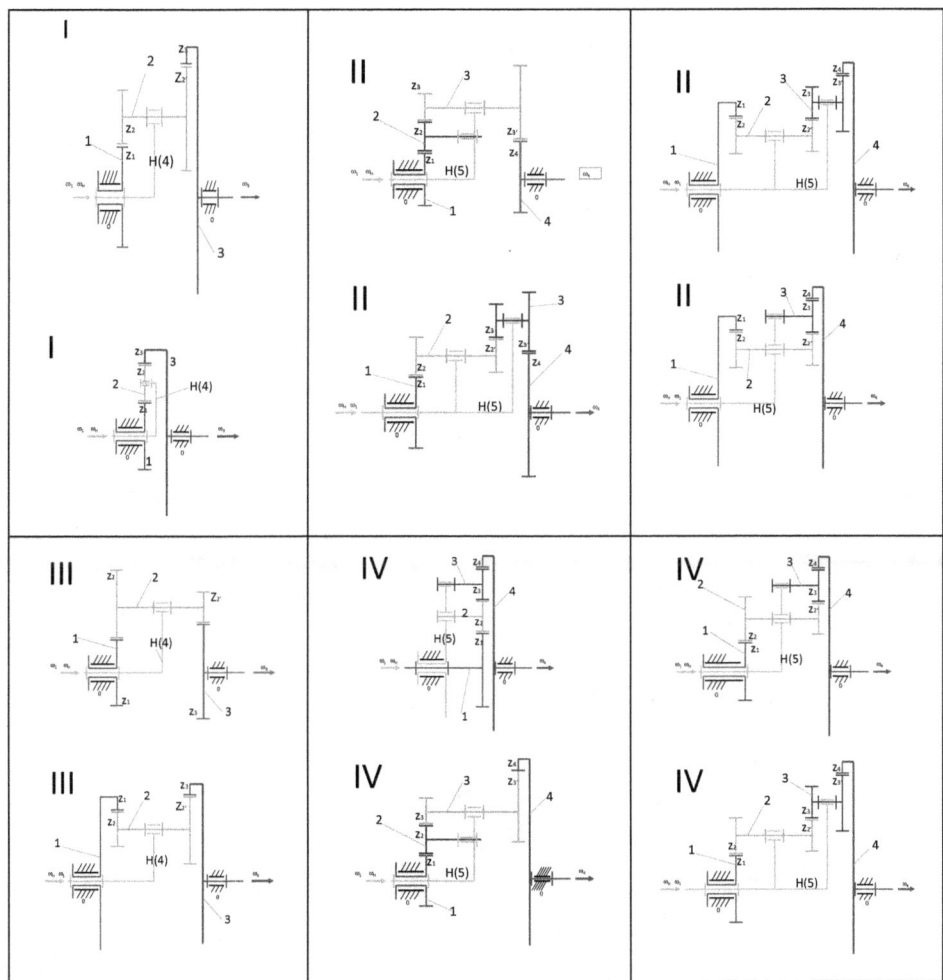

Fig. 3. Planetary systems

$$\begin{cases}
i_{13}^{H} = -\dfrac{z_2}{z_1} \cdot \dfrac{z_3}{z_{2'}} & \text{for} \quad I \quad \text{of} \quad up \\[3mm]
i_{13}^{H} = -\dfrac{z_3}{z_1} & \text{for} \quad I \quad \text{of} \quad down \\[3mm]
i_{13}^{H} = \dfrac{z_2}{z_1} \cdot \dfrac{z_3}{z_{2'}} & \text{for} \quad III \quad \text{of} \quad up \\[3mm]
i_{13}^{H} = \dfrac{z_2}{z_1} \cdot \dfrac{z_3}{z_{2'}} & \text{for} \quad III \quad \text{of} \quad down \\[3mm]
i_{14}^{H} = -\dfrac{z_3}{z_1} \cdot \dfrac{z_4}{z_{3'}} & \text{for} \quad II \quad \text{left} \quad up \\[3mm]
i_{14}^{H} = -\dfrac{z_2}{z_1} \cdot \dfrac{z_3}{z_{2'}} \cdot \dfrac{z_4}{z_{3'}} & \text{for} \quad II \quad \text{right} \quad up \\[3mm]
i_{14}^{H} = -\dfrac{z_2}{z_1} \cdot \dfrac{z_3}{z_{2'}} \cdot \dfrac{z_4}{z_{3'}} & \text{for} \quad II \quad \text{left} \quad down \\[3mm]
i_{14}^{H} = -\dfrac{z_2}{z_1} \cdot \dfrac{z_4}{z_{2'}} & \text{for} \quad II \quad \text{right} \quad down \\[3mm]
i_{14}^{H} = \dfrac{z_4}{z_1} & \text{for} \quad IV \quad \text{left} \quad up \\[3mm]
i_{14}^{H} = \dfrac{z_2}{z_1} \cdot \dfrac{z_4}{z_{2'}} & \text{for} \quad IV \quad \text{right} \quad up \\[3mm]
i_{14}^{H} = \dfrac{z_3}{z_1} \cdot \dfrac{z_4}{z_{3'}} & \text{for} \quad IV \quad \text{left} \quad down \\[3mm]
i_{14}^{H} = \dfrac{z_2}{z_1} \cdot \dfrac{z_3}{z_{2'}} \cdot \dfrac{z_4}{z_{3'}} & \text{for} \quad IV \quad \text{right} \quad down
\end{cases} \tag{4}$$

The planetary mechanisms are less synthesized based by their mechanical efficiency developed in operation, although this criterion is part of the real dynamics of mechanisms, being also the most important criterion in terms of performance of a mechanism.

Even when it used the efficiency criterion, the determination of the planetary yield, is made only with approximate relationships.

The most widely recognized method is one method of Russian school of mechanisms.

This chapter determines the real efficiency of the planetary trains (the exact method).

The automatic transmissions have been added slowly from airplanes to automobiles, and were then generalized to various vehicles. By using the formulas indicated in this paper for calculating the dynamic of planetary mechanisms, planetary trains, and planetary systems, used in aircraft and vehicles, their automatic transmissions can be achieved better than those known today.

In this mode it is resolving one important problem of the dynamics of planetary mechanisms. For a complete dynamic, may need to be determined and the dynamics deformation mechanisms (see the Figure 4). But this is not part of the subject matter of this book.

Fig. 4. *The deformation of a planetary mechanism*

3. Dynamic synthesis, based on performance achieved

Dynamic synthesis of planetary trains based on performance achieved, can be made with the relationships presented below.

For a normally planetary system (Figure 2) the mechanical efficiency is determined by starting from the relationship 5, which gives the lost power (P_l) in function of the input power (P_H) and the output power (P_3 or P_4, and generic P_f).

$$
\begin{aligned}
P_l &= P_H - P_3 = M_H \cdot \omega_H - M_3 \cdot \omega_3 = \\
&= (M_3 + M_1) \cdot \omega_H - M_3 \cdot \omega_3 = \\
&= M_3 \cdot \omega_H - M_3 \cdot \omega_3 + M_1 \cdot \omega_H = \\
&= M_3 \cdot (\omega_H - \omega_3) + M_1 \cdot \omega_H
\end{aligned}
\tag{5}
$$

It is known the Willis relationship (6), from that it can explicit the moment (M_1). With M_1 put in the relationship (5) it obtains the expression (7).

$$
\begin{cases}
\eta_{13}^H = \dfrac{P_3^H}{P_1^H} = \dfrac{M_3 \cdot \omega_3^H}{M_1 \cdot \omega_1^H} = \dfrac{M_3 \cdot (\omega_3 - \omega_H)}{M_1 \cdot (\omega_1 - \omega_H)} = \\[2mm]
= \dfrac{M_3}{M_1} \cdot \dfrac{\omega_3 - \omega_H}{-\omega_H} = \dfrac{M_3}{M_1} \cdot \left(1 - \dfrac{\omega_3}{\omega_H}\right) = \\[2mm]
= \dfrac{M_3}{M_1} \cdot (1 - i_{3H}) = \dfrac{M_3}{M_1} \cdot (1 - i_{3H}^1) \Rightarrow \\[2mm]
\Rightarrow M_1 = \dfrac{M_3}{\eta_{13}^H} \cdot (1 - i_{3H}^1)
\end{cases}
\tag{6}
$$

$$
\begin{cases}
P_1 = M_3 \cdot (\omega_H - \omega_3) + M_1 \cdot \omega_H = \\[2mm]
= M_3 \cdot (\omega_H - \omega_3) + \dfrac{M_3 \cdot \omega_H}{\eta_{13}^H} \cdot (1 - i_{3H}) = \\[2mm]
= M_3 \cdot \omega_3 \cdot \left(\dfrac{\omega_H}{\omega_3} - 1\right) + M_3 \cdot \omega_3 \cdot \left(\dfrac{\omega_H}{\omega_3} - 1\right) \cdot \dfrac{1}{\eta_{13}^H} = \\[2mm]
= M_3 \cdot \omega_3 \cdot \left(\dfrac{\omega_H}{\omega_3} - 1\right) \cdot \left(1 + \dfrac{1}{\eta_{13}^H}\right) = \\[2mm]
= M_3 \cdot \omega_3 \cdot (i_{H3} - 1) \cdot \dfrac{1 + \eta_{13}^H}{\eta_{13}^H} = P_3 \cdot (i_{H3} - 1) \cdot \dfrac{1 + \eta_{13}^H}{\eta_{13}^H} \\[2mm]
\Rightarrow P_p = |P_1| = P_3 \cdot \dfrac{1 + \eta_{13}^H}{\eta_{13}^H} \cdot |i_{H3} - 1|
\end{cases}
\tag{7}
$$

The exact efficiency can be obtained by putting the expression of the "real lost power Pp" (the absolute lost power, determined by the relationship 7) in the formula of the mechanical efficiency of one planetary system (8).

$$\begin{cases} \eta_{H3}^1 = \dfrac{P_3}{P_H} = \dfrac{P_3}{P_3 + P_p} = \dfrac{P_3}{P_3 + P_3 \cdot \dfrac{1+\eta_{13}^H}{\eta_{13}^H} \cdot \left| i_{H3} - 1 \right|} = \\[2em] = \dfrac{1}{1 + \dfrac{1+\eta_{13}^H}{\eta_{13}^H} \cdot \left| i_{H3} - 1 \right|} = \dfrac{1}{1 + \dfrac{1+\eta_{13}^H}{\eta_{13}^H} \cdot \left| i_{H3}^1 - 1 \right|} \end{cases} \qquad (8)$$

For the mechanisms with four toothed wheels the efficiency takes the form (9).

$$\begin{cases} \eta_{H4}^1 = \dfrac{P_4}{P_H} = \dfrac{P_4}{P_4 + P_p} = \dfrac{P_4}{P_4 + P_4 \cdot \dfrac{1+\eta_{14}^H}{\eta_{14}^H} \cdot \left| i_{H4} - 1 \right|} = \\[2em] = \dfrac{1}{1 + \dfrac{1+\eta_{14}^H}{\eta_{14}^H} \cdot \left| i_{H4} - 1 \right|} = \dfrac{1}{1 + \dfrac{1+\eta_{14}^H}{\eta_{14}^H} \cdot \left| i_{H4}^1 - 1 \right|} \end{cases} \qquad (9)$$

4. Conclusions

The planetary system efficiency given by the exact formula is less than the calculated efficiency by known classical formulas.

If these new relationships are correct, the planetary systems work generally with lower efficiency.

References

[1] Petrescu, F.I., Petrescu, R.V., *Planetary Trains*, Book, LULU Publisher, USA, April 2011, ISBN 978-1-4476-0696-3, pages 204.

Annex

A brief history about the emergence and evolution of gearing mechanisms

A brief history about the emergence and evolution of gearing mechanisms

Top of the use of sprocket mechanisms must be sought in ancient Egypt with at least a thousand years before Christ. Here were used for the first time, transmissions wheeled "spurred" to irrigate crops and worm gears to the cotton processing.

With 230 years BC, in the city of Alexandria in Egypt, they have been used the wheel with more levers and gear rack.

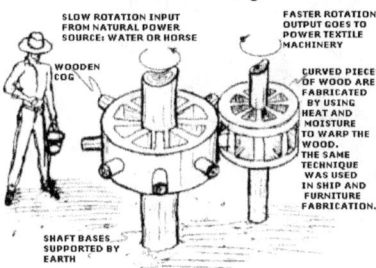

Such gears have been constructed and used beginning from the earliest times, to the top for lifting the heavy anchors of vessels and for claim catapults used on the battlefields. Then, they were introduced in cars with wind and water (as a reducing or multiplying at the pump from windmills or water).

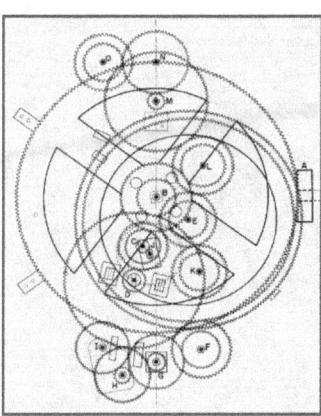

The Antikythera Mechanism is the name given to an astronomical calculating device, measuring about 32 by 16 by 10 cm, which was discovered in 1900 in a sunken ship just off the coast of Antikythera, an island between Crete and the Greek mainland. Several kinds of evidence point incontrovertibly to around 80 B.C. for the date of the shipwreck. The device, made of bronze gears fitted in a wooden case, was crushed in the wreck, and parts of the faces were lost, "the rest then being coated with a hard calcareous deposit at the same time as the metal corroded away to a thin core coated with hard metallic salts preserving much of the former shape of the bronze" during the almost 2000 years it lay submerged. (Antikythera 1).

It is hard to exaggerate the singularity of this device, or its importance in forcing a complete re-evaluation of what had been believed about technology in the ancient world. For this box contained some 32 gears, assembled into a mechanism that accurately reproduced the motion of the sun and the moon against the background of fixed stars, with a differential giving their relative position and hence the phases of the moon.

Modern adventure began with the gear wheel spurred of Leonardo da Vinci, in the fifteenth century. He founded the new kinematics and dynamics stating inter alia the principle of superposition of independent movements.

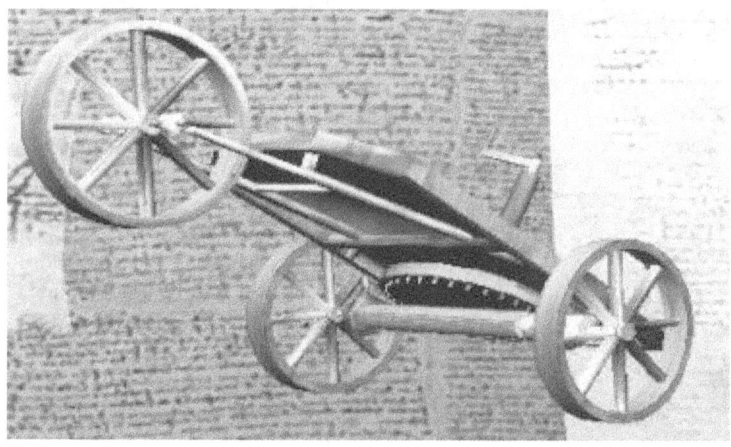

LEO's spurred wheel presented in detail. This CAR driven by a crank, is constructed simply, having a transmission consisting of a grooved wheel spurs and an axle. Spurs work on grooves, rotating axle. For the movement to be able to transmit, the axle has a groove in to a side.

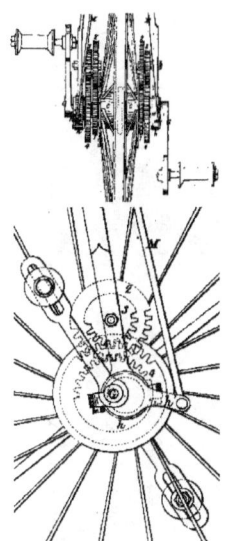

Benz had engine with transmissions sprocket gearing and Gear chain (patented after 1882). On the right (up and down), you can see the drawings of a patent first gear transmission (first gearing patent) and of gearing wheels with chain made in 1870 by the British **Starley & Hillman.**

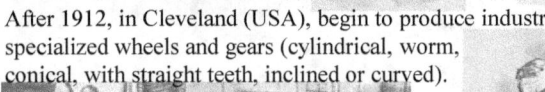

After 1912, in Cleveland (USA), begin to produce industrial specialized wheels and gears (cylindrical, worm, conical, with straight teeth, inclined or curved).

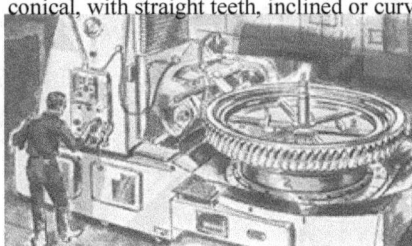

The old mechanisms with gearing (and bars) that were preserved:
A-ratchet;
B-worm screw mechanism and worm;
C-pendulum;
D-Leonardo da Vinci mechanism with worm, crank, rods and flies;
E-Planetary gears.

Gearing today.

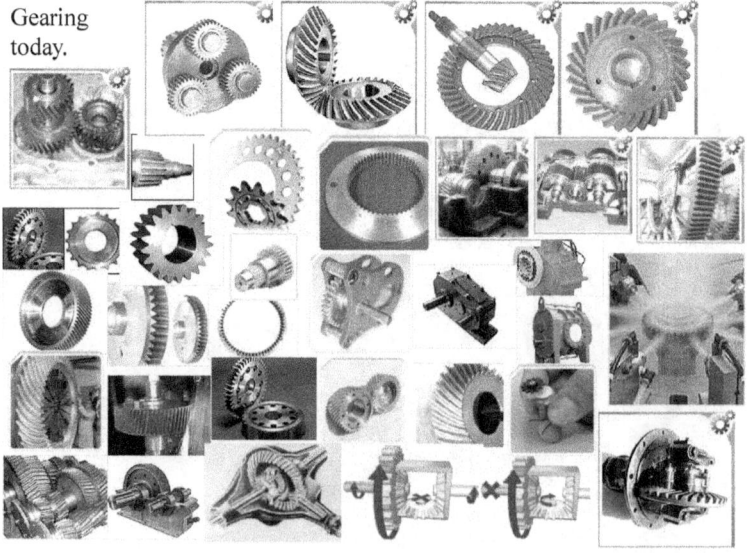

Gears and gearing
for
heavy machinery.

Specialized gear reducers used in:

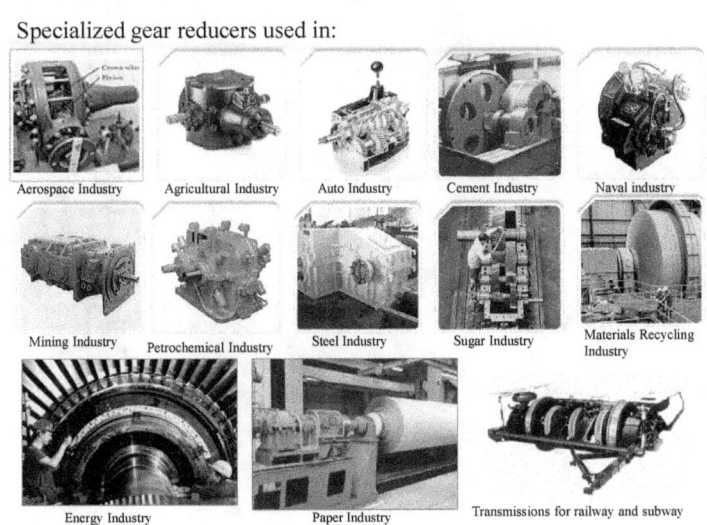

Aerospace Industry | Agricultural Industry | Auto Industry | Cement Industry | Naval industry

Mining Industry | Petrochemical Industry | Steel Industry | Sugar Industry | Materials Recycling Industry

Energy Industry | Paper Industry | Transmissions for railway and subway

Some areas of use gearing

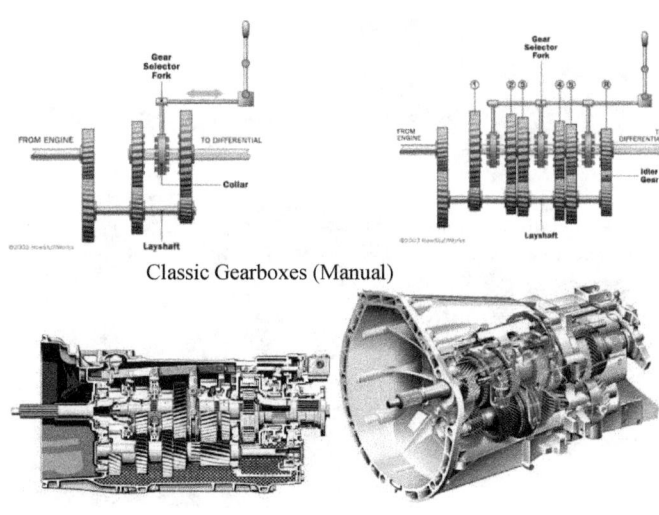

Classic Gearboxes (Manual)

The first automatic transmission
(gears and planetary gears)

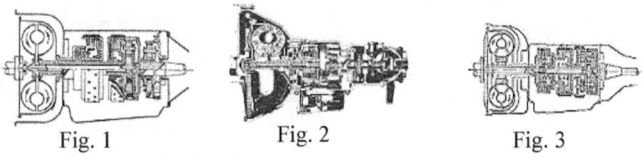

Fig. 1   Fig. 2   Fig. 3

Hydra-Matic transmission in Fig. 1, is an experienced model automatic transmission model Oldsmobile General Motors Corporation in 1940; The Dynaflow (Fig. 2), also created in 1948 by General Motors for Buick, was more effective;

Powerglide particular model, which was designed by General Motors in 1953, was a typical two-speed automatic transmission, which served as a standard model for other companies, so based on him, Ford's makes the Ford-O-Matic model (Fig. 3).

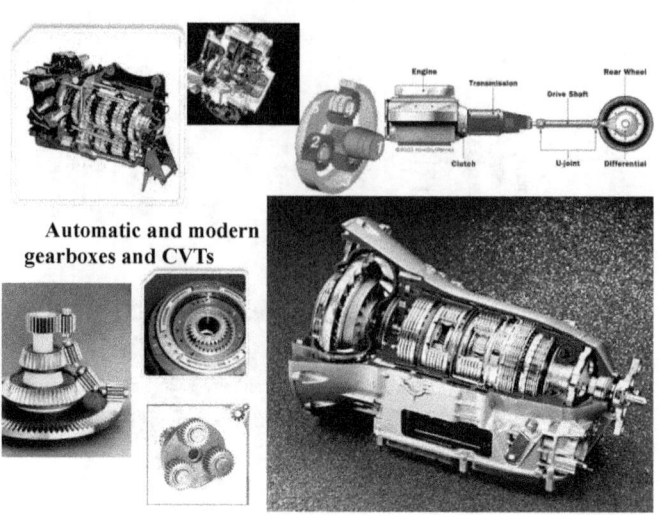

Automatic and modern gearboxes and CVTs

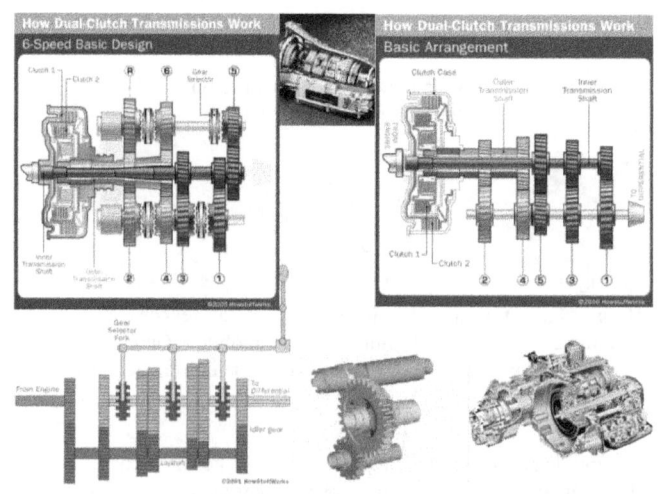

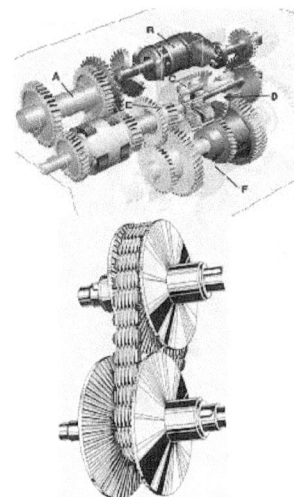

News
CVT classical or Variable Transmission (automatics) continuous (the classical model).

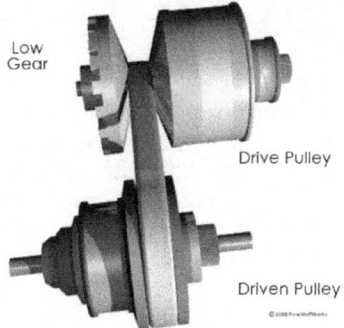

Low Gear

Drive Pulley

Driven Pulley

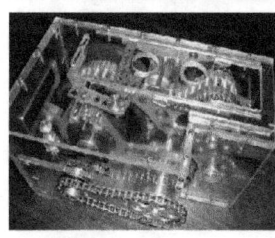

Gearbox exchanger swing mechanism. This mechanism, oscillating gear shifting is driven by a rotary cylinder, coupled to a lock mechanism (result is a high-speed gear shift). Oscillatory mechanism can be seen in next slide.

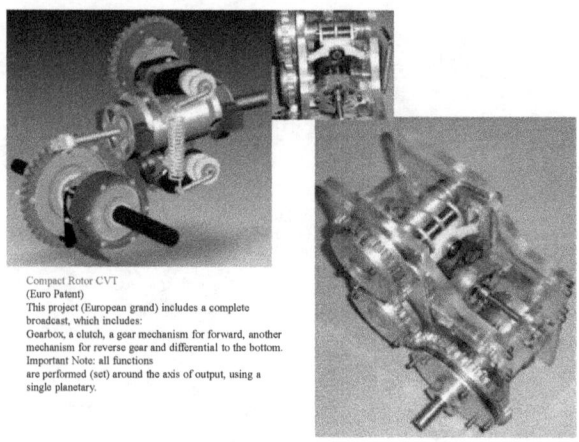

Compact Rotor CVT
(Euro Patent)
This project (European grand) includes a complete
broadcast, which includes:
Gearbox, a clutch, a gear mechanism for forward, another
mechanism for reverse gear and differential to the bottiom.
Important Note: all functions
are performed (set) around the axis of output, using a
single planetary.

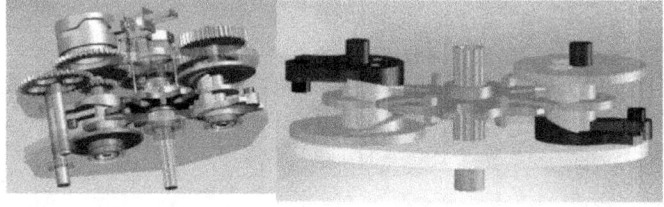

This patented mechanism of action for
exchangers speed sequential mechanism is based on a cross of Malta,
who works as a factor acting as the leader position for the gear change
soon.
This mechanism has the advantage of eliminating the command and the
hydraulic or servo, control, actuation and timing are all making with
Maltese cross mechanism changed.
The mechanism of action can be seen in next slide.

Actuator, the cross of Malta, amended.

This compact model of CVT (continuously variable transmission), was
presented at the "Frankfurt motor show, September 2007 and
represents a fully functional model continuously variable transmission,
compact, and autonomous. This model was exhibited at the exhibition
organized and otherwise in the "SAE Commercial Vehicle Engineering
Congress & Exhibition, October 30 - November 1, 2007 Rosemont
(Chicago), Illinois, USA. Variable transmission mechanism is based on
a drum, chain and bar.

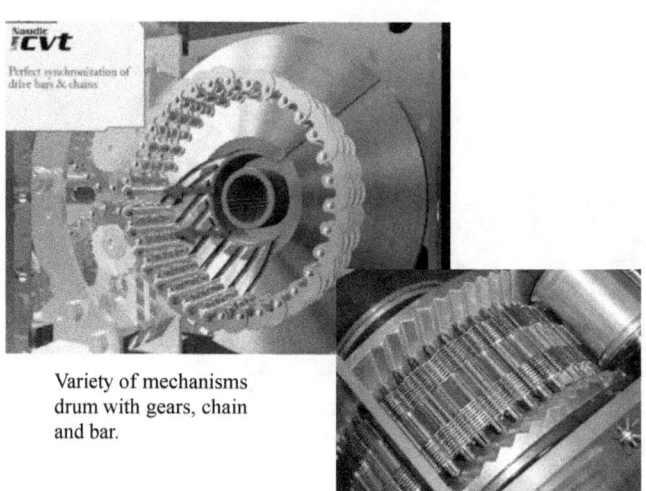

Variety of mechanisms
drum with gears, chain
and bar.

SEE YOU SOON!